Digital Da Vinci

Newton Lee
Editor

Digital Da Vinci

Computers in the Arts and Sciences

 Springer

Editor
Newton Lee
Newton Lee Laboratories, LLC
Tujunga
California
USA

ISBN 978-1-4939-4915-1   ISBN 978-1-4939-0965-0 (eBook)
DOI 10.1007/978-1-4939-0965-0
Springer New York Heidelberg Dordrecht London

Printed on acid-free paper

Springer is part of Springer Science+Business Media (www.springer.com)

"Study the science of art and the art of science."

Leonardo da Vinci

About the Book

"Science is art," said Regina Dugan, senior executive at Google and former director of DARPA. "It is the process of creating something that never exists before. ... It makes us ask new questions about ourselves, others; about ethics, the future." In *Computers in the Arts and Sciences*—the second volume of the *Digital Da Vinci* book series—Newton Lee leads the discussions on the world's first computer art in the 1950s and the actualization of *Star Trek's* holodeck in the future with the help of artificial intelligence and cyborgs.

In this book, Gavin Sade describes experimental creative practices that bring together arts, science, and technology in imaginative ways; Mine Özkar expounds visual computation for good designs based on repetition and variation; Raffaella Folgieri, Claudio Lucchiari, Marco Granato, and Daniele Grechi introduce Brain Art, a brain-computer interface that allows users to create drawings using their own cerebral rhythms; Nathan Cohen explores artificially created spaces that enhance spatial awareness and challenge our perception of what we encounter; Keith Armstrong discusses embodied experiences that affect the mind and body of participating audiences; Diomidis Spinellis uses Etoys and Squeak in a scientific experiment to teach the concept of physical computing; Benjamin Cowley explains the massively multiplayer online game "Green My Place" aimed at achieving behavior transformation in energy awareness; Robert Niewiadomski and Dennis Anderson portray 3-D manufacturing as the beginning of common creativity revolution; Stephen Barrass takes 3-D printing to another dimension by fabricating an object from a sound recording; Mari Velonaki examines the element of surprise and touch sensing in human-robot interaction; and Roman Danylak surveys the media machines in light of Marshall McLuhan's dictum "the medium is the message."

Digital Da Vinci: Computers in the Arts and Sciences is dedicated to polymathic education and interdisciplinary studies in the digital age empowered by computer science. Educators and researchers ought to encourage the new generation of scholars to become as well rounded as a Renaissance man or woman.

Contents

**11 Human-Robot Interaction in Prepared Environments:
Introducing an Element of Surprise by Reassigning Identities
in Familiar Objects**.. 211
Mari Velonaki

12 The Messages of Media Machines: Man-Machine Symbiosis............... 223
Roman Danylak

Contributors

Dennis Anderson St. Francis College, Brooklyn, USA

Keith Armstrong QUT Creative Industries, Brisbane, Australia

Stephen Barrass University of Canberra, Bruce, Australia

Nathan Cohen University of the Arts London, London, UK

Benjamin Cowley Cognitive Brain Research Unit, Cognitive Science, Institute of Behavioural Sciences, University of Helsinki, Helsinki, Finland.

Brain and Work Research Centre, Finnish Institute of Occupational Health, Helsinki, Finland

Roman Danylak University of Technology, Sydney, Ultimo, Australia

Raffaella Folgieri Department of Economics, Management and quantitative Methods, Università degli Studi di Milano, Milan, Italy

Marco Granato Department of Health Sciences, Università degli Studi di Milano, Milan, Italy

Daniele Grechi Department of Naval, Electrical, Electronic and Telecommunications Engineering, Polytechnical School, Università degli Studi di Genova, Genoa, Italy

Newton Lee Newton Lee Laboratories, LCC, Tujunga, CA, USA

School of Media, Culture & Design, Woodbury University, Burbank, USA

Claudio Lucchiari Department of Health Sciences, Università degli Studi di Milano, Milan, Italy

Robert Niewiadomski New York City Department of Education (NYCDOE) and Teach For America, New York, USA

Mine Özkar Istanbul Technical University, Istanbul, Turkey

Gavin Sade School of Design, Queensland University of Technology, Brisbane, Australia

Diomidis Spinellis Department of Management Science and Technology, Athens University of Economics and Business, Athens, Greece

Mari Velonaki University of New South Wales, Kensington, Australia

About the Authors

Dennis Anderson is Chair and Professor of Management and Information Technology at St. Francis College. Prior to this appointment he was a Professor of Information Systems & Computer Science and served as Associate Dean at Pace University. He is a strong advocate of technology-enhanced learning, emerging technologies, sustainable technologies, and knowledge entrepreneurship. He also has taught at NYU, City University of New York, and Pace University. Dennis received his Ph.D., M.Phil. and Ed.M. from Columbia University. In addition, he holds an M.S. in Computer Science from NYU's Courant Institute of Mathematical Sciences.

Keith Armstrong has specialized for 18 years in collaborative, hybrid, new media works with an emphasis on innovative performance forms, site-specific electronic arts, networked interactive installations, alternative interfaces, public arts practices and art-science collaborations. His ongoing research focuses on how scientific and philosophical ecologies can both influence and direct the design and conception of networked, interactive media artworks. Keith's artworks have been shown and profiled extensively both in Australia and overseas and he has been the recipient of numerous grants from the public and private sectors. He was formerly an Australia Council New Media Arts Fellow, a doctoral and Postdoctoral New Media Fellow at QUT's Creative Industries Faculty and a lead researcher at the ACID Australasian Cooperative Research Centre for Interaction Design. He is currently a part-time Senior Research Fellow (2 days pw.) at QUT and an actively practicing freelance new media artist.

Stephen Barrass is a researcher and academic at the University of Canberra where he lectures in Digital Design and Media Arts in the Faculty of Arts and Design. He holds a B.E. in Electrical Engineering from the University of New South Wales (1986) and a Ph.D. titled Auditory Information Design from the Australian National University (1997). He was a Post-Doctoral Fellow at the Fraunhofer Institute for Media Kommunication in Bonn (1998) and Guest Researcher in Sound Design and Perception at IRCAM in Paris (2009).

Nathan Cohen is a professional artist exhibiting internationally for over 25 years, including solo shows at Annely Juda Fine Art, London; Museum Mondriaanhuis, Holland; Tokyo Gallery, Japan and many other venues worldwide. His interdisciplinary research in art and science embraces neuroscience, optics and augmented reality technologies resulting in recent interactive art installations exhibited at Ars *Electronica*, Austria (*Hybrid Ego* 2008), the Aisho Miura gallery (*Intangible Spaces* 2010), Japan and University College London (*Another Way of Seeing* 2013). In collaboration with Tachi Lab (Tokyo) and researchers in Japan and the UK he creates artworks that challenge spatial perception, incorporating motion sensing and real-time projection into 2 and 3-dimensional constructions created to give the impression of multi-layered spaces. In 2011 Nathan Cohen established the first Masters program in Art and Science (Central Saint Martins, University of the Arts London) and is currently Director. His professional activities also embrace publishing, directing an archive, curating exhibitions internationally and writing. He was a recipient of the *Vordemberge-Gildewart* Award in 1994 and studied at the Slade School of Fine Art, UCL (BA Hons.) and Chelsea School of Art (MA), London.

Benjamin Cowley received his Bachelors degree on Information and Communications Technology from Trinity College Dublin, Ireland, in 2003, and subsequently defended his PhD in Computer Science at the University of Ulster, Northern Ireland, in 2009. Initial post-doctoral projects focused on investigating the psychophysiological correlates of learning in the domain of serious games, in the Centre for Knowledge Innovation and Research at Aalto University, Helsinki. Presently he studies neurofeedback games for attentional disorder therapy at the University of Helsinki. Research interests are in games for learning, cognitive science, and attention as a component of positive psychology.

Roman Danylak is an interactive artist. He completed a Ph.D. at the Creativity and Cognition Studios, University of Technology, Sydney in 2008, specializing in design for gesture and emotions using semiotics. His work, *To be or not to be*, was featured at Sydney's Powerhouse Museum. He has presented at numerous international academic conferences in Sweden, Japan, USA, Italy, UAE, France and Germany. He has lectured and designed online curriculum for Stockholm University in Interactive Art. His work began with Metamorph (1996), a prototype of interactive performance, and was the one of the first Australian works to be featured on the World Wide Web. He has also worked for the Australian National Playwrights' Centre developing scripts to professional performance level and has published many critical reviews on art and design. As an artist he has worked in film, TV and theatre as writer, musician and performer.

Raffaella Folgieri, Ph.D. in Computer Science, is Assistant Professor in Computer Skills at the Faculty of Political Science and of Information Technology at the Faculty of Political Science and at the Faculty of Medicine (Medical and Pharmaceutical Biotechnologies). She also teaches Information Technology Representation of Knowledge in the post-degree course in Cognitive Science and Decision Making, Virtual Reality in the Information Technology and Digital Communication degree course and Project Management at the Faculty of Mathematics, Physics, and Natural Sciences of the University of Milan. Member of Italian Society of Engineering and of SIREN (Italian Neural Networks Society), she has published her research in several journal articles (main fields of interests: Brainomics; Brain Computer Interfaces; Virtual Reality; Bioinformatics; Machine Learning and AI; Quality assessment in complex software development; e-learning). Her work explores some of the central issues in cognitive research such as how people move from skilledperformance to problem solving, how a person learns, manages errors, interprets visual stimuli, and communicates. She coordinates the research group Beside, focused on interpersonal, machine-machine and brain-machine communication mediated by technology, and ExCog (jointly with Prof. Lucchiari), aiming to study MIND in all its complexity and all possible shapes.

Marco Granato is currently a post-degree student in Cognitive Science. He has a degree in Computer Science from the University of Milan. He is a member of the research groups Beside and ExCog (Extended Cognition) at the University of Milan, under the supervision of Prof. Folgieri. His research interests cover applied research on the brain and computer interaction.

Daniele Grechi is a research fellow in the Department of Naval, Electrical, Electronic and Telecommunications Engineering of the Polytechnical School, at the University of Genoa. He received a Master Degree in Banking and Finance from the University of Insubria (Varese) and he worked for the IUSS of Pavia for a software banking risk project. He is also contract professor of Statistics in the Department of Economics of the University of Insubria. His current research includes software engineering, software metrics, software development methodology and statistical software analysis.

Newton Lee is founding director of the Woodbury University Digital Media Lab and adjunct professor of Media Technology at the School of Media, Culture & Design. He is also CEO of Newton Lee Laboratories LLC, president of the Institute for Education, Research, and Scholarships, and founding editor-in-chief of ACM Computers in Entertainment. Previously, he was a research scientist at AT&T Bell Laboratories, senior producer and engineer at The Walt Disney Company, research staff member at the Institute for Defense Analyses, and research scientist at Virginia Tech Library Systems. Lee graduated Summa Cum Laude from Virginia Tech with a B.S. and M.S. degree in Computer Science, and he earned a perfect GPA from Vincennes University with an A.S. degree in Electrical Engineering and an honorary doctorate in Computer Science. He is the co-author of *Disney Stories: Getting to Digital*; the author of the Total Information Awareness book series including *Facebook Nation and Counterterrorism and Cybersecurity;* and the editor of the Digital Da Vinci book series including *Computers in Music and Computers in the Arts and Sciences.*

Claudio Lucchiari is Assistant professor in Cognitive Pscyhology, University of Milan. He has a degree in psychology and a Ph.D. in Communication Psychology. He was tutor and professor of cognitive psychology at the Catholic University of Milan and the University of Urbino (2003–2006). He worked as neuro-psychologist and cognitive psychologist at the Neurological National Institute of Milan (2002–2006), where he focused on the neurological basis of thought, medical decision making, shared decisions, doctor-patient communication and health related quality of life. Since 2006, he has been a lecturer at the University of Milan and a member of the IRIDe (Interdisciplinary Research and Intervention on Decision) research centre. His research activities focus on decision making and the application of cognitive science in various fields. In particular, he conducts research on medical decision making, psychoeconomics, neuroeconomics (risk perception, emotional and neuro-marketing), the neural correlates of decisions and creativity. Furthermore he is developing a BCI-based neuro-cognitive program aimed at empowering cognitive performances.

Robert Niewiadomski is an educator with the NYCDOE and a Teach For America corps member. He earned a B.A in Philosophy from Columbia University and is currently pursuing graduate studies both at Columbia and Fordham University. At Columbia, he investigated ethical aspects of gene patenting and presented at several philosophy conferences. Prior to his arrival at Columbia, he studied at the University of Warsaw and participated in the European Union sponsored Youth Program in the UK. Robert's main areas of interest include philosophy of science as well as ethical aspects of medical and technological advances.

Mine Özkar is a visiting professor at MIT for Spring 2013 and an associate professor of architecture at Istanbul Technical University, where she also serves on the executive committee for the Program in Computational Design. She earned her SMArchS in design inquiry and her PhD in design and computation from MIT. In some of her previous work, she has interpreted the history and theory of progressive pedagogy in art and design from a computational perspective. Her current research focuses on shape representation, spatial computation, and computational design methods. She also publishes on the theory and practice of foundational design education and the ongoing global and local curriculum reforms in architectural education. She recently guest-edited the 2012 ACM SIGGRAPH/Leonardo ISAST Special Issue and co-edited a book titled *Shaping Design Teaching*.

Gavin Sade is an artist, designer and researcher in the field of interactive computational media, with a background in music and sonology. He is currently the Head of Interactive and Visual Design in the Creative Industries Faculty at the Queensland University of Technology. Gavin holds a Bachelor of Music (Sonology) from the Queensland Conservatorium of Music, and a PhD in interactive media arts and sustainability. He has been creating interactive media systems and electronic art since 1990 when he began working with electronic music group Vision 4/5 and later with Keith Armstrong and the Transmute Collective. In 2003 he formed Kuuki (http://kuuki.com.au), a creative media collective, and has since lead the production of a number of high profile electronic artworks that have been exhibited in international settings, such as the 2011 International Symposium of Electronic Art in Istanbul and the Museum of Contemporary Art Taipei. Gavin's practice is guided by an ecological philosophy inspired by vegetarianism and the critical design philosophy of defuturing, and is a mix of electronic art and critical design.

Diomidis D. Spinellis is a professor at the Department of Management Science and Technology at the Athens University of Economics and Business in Greece. Previously, he was the General Secretary of Information Systems at the Greek Ministry of Finance. He is a member of the IEEE Software editorial board, contributing the Tools of the Trade column. He is a four-time winner of the International Obfuscated C Code Contest. Spinellis holds a Master of Engineering degree in Software Engineering and a Ph.D. in Computer Science from Imperial College London. He has authored and coauthored *Code Reading, Code Quality,* and *Beautiful Architecture.*

Mari Velonaki has worked as a researcher in the field of interactive installation art since 1997. Mari has created interactive installations that incorporate movement, speech, touch, breath, electrostatic charge, artificial vision and robotics. In 2003 Mari's practice expanded to robotics, when she initiated and led a major Australian Research Council art/science research project 'Fish–Bird: Autonomous Interactions in a Contemporary Arts Setting 2004–2007' in collaboration with roboticists at the Australian Centre for Field Robotics. In 2006 she co- founded, with David Rye, the Centre for Social Robotics, a centre dedicated to inter- disciplinary research into human-robot interaction in spaces that incorporate the general public. In 2007 Mari was awarded an Australia Council for the Arts Visual Arts Fellowship. In 2009 she was awarded an Australia Research Council Queen Elizabeth II Fellowship (2009–2013) for the creation of a new robot in order to develop an understanding of the physicality that is possible and acceptable between a human and a robot. Mari is currently an Associate Professor and the director of a new lab, the Creative Robotics Lab, at the National Institute of Experimental Arts at the College of Fine Arts, The University of New South Wales. Mari's artworks have been exhibited worldwide.

Chapter 1
From a Pin-up Girl to Star Trek's Holodeck: Artificial Intelligence and Cyborgs

Newton Lee

"Science is art. It is the process of creating something that never exists before....
It makes us ask new questions about ourselves, others; about ethics, the future."–
Regina Dugan, senior executive at Google and former director of DARPA (Denise
2013).

1 The World's First Computer Art: A Pin-Up Girl

Sometime between 1956 and 1958, an anonymous IBM programmer rendered a
glowing image of a pin-up girl on a cathode ray tube screen of a $ 238 million U.S.
military computer at Fort Lee, Virginia. "The pin-up image itself was programmed
as a series of short lines, or vectors, encoded on a stack of about 97 Hollerith type
punched cards," recalled Airman First Class Lawrence A. Tipton who took the Po-
laroid photo shown in Fig. 1.1 that somewhat resembles a hybrid of Betty Boop and
Esquire's December 1956 calendar pin-up by George Petty (Benj 2013).

A few years later in 1963, Ivan Sutherland developed a computer program called
Sketchpad for his Ph.D. dissertation at MIT. Sketchpad allowed users to create and
manipulate graphic images on a CRT screen using a light pen, and to store the objects for future editing. He described himself as a visual thinker: "If I can picture
possible solutions, I have a much better chance of finding the right one" (Burton
1988). For his pioneering and visionary contributions to computer graphics, Sutherland received the A.M. Turing Award in 1988.

N. Lee (✉)
Newton Lee Laboratories, LLC, Tujunga, CA, USA
e-mail: newton@newtonlee.com

School of Media, Culture & Design, Woodbury University, Burbank, CA, USA
e-mail: newton.lee@woodbury.edu

N. Lee (ed.), *Digital Da Vinci,* DOI 10.1007/978-1-4939-0965-0_1,
© Springer Science+Business Media New York 2014

Fig. 1.1 A pin-up girl photographed by Airman First
Class Lawrence A. Tipton.
(Courtesy of SMECC/South-
west Museum of Engineer-
ing, Communications and
Computation)

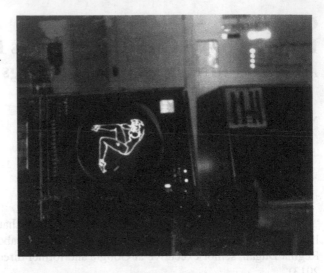

Fig. 1.1 A pin-up girl photographed by Airman First Class Lawrence A. Tipton. (Courtesy of SMECC/Southwest Museum of Engineering, Communications and Computation)

2 Artificial Intelligence (A.I.): The Turing Test

The A.M. Turing Award, sometimes referred to as the "Nobel Prize" of Computing, is an annual prize given by the Association for Computing Machinery (ACM) to individuals for their "major contributions of lasting importance to computing" (Association for Computing Machinery 2012). The award is also accompanied by a cash prize of $ 250,000, which in recent years has been underwritten by Intel and Google.

Alan Mathison (A.M.) Turing was a British mathematician, cryptanalyst, and computer scientist who is widely considered to be the father of computer science and artificial intelligence. In 1950, Alan Turing introduced the famous imitation game (aka the Turing Test) as a test of a machine's ability to exhibit intelligent behavior equivalent to, or indistinguishable from, that of a human (Turing 1950).

John McCarthy coined the term "artificial intelligence" (A.I.) a few years later in 1955 when he proposed a summer research conference: "The study is to proceed on the basis of the conjecture that every aspect of learning or any other feature of intelligence can in principle be so precisely described that a machine can be made to simulate it" (Myers 2011).

Joseph Weizenbaum's ELIZA program in 1966 was arguably a successful attempt to pass the Turing Test. ELIZA simulated a conversation between a patient and a psychotherapist by using a person's responses to shape the computer's replies (Weizenbaum 1966). Although ELIZA has no understanding of natural language or psychotherapy, many people were fooled into believing that ELIZA was human. More recently in 2010, A.I. researcher Rollo Carpenter developed the Cleverbot chat engine that learns from a growing database of 20+ million human online conversations (Saenz 2010). Cleverbot passed the Turing Test with flying colors at the 2011 Techniche festival in Guwahati, India. Chat participants and the audience rated the humanness of all chat responses, with Cleverbot voted 59.3% human, while the humans themselves were rated just 63.3% human (Aron 2011).

But A.I. is much more useful than being able to fool people.

3   A.I. and Expert Systems: From Chemistry To Chess To Jeopardy!

An expert system is an A.I. program that emulates the decision-making ability of a human expert by reasoning about knowledge.

In 1965, A.I. researcher Edward Feigenbaum and geneticist Joshua Lederberg of Stanford University began to develop DENDRAL, a chemical-analysis expert system that hypothesizes a test substance's molecular structure (Copeland 2013). DENDRAL's performance has rivaled that of chemists.

In the early 1970s at Stanford University, Edward Shortliffe created MYCIN, a rule-based expert system that consults with physicians about the diagnosis and treatment of infectious diseases (Association for Computing Machinery 2013). MYCIN has been shown to outperform doctors in some cases.

For my undergraduate research at Virginia Tech in 1980s, I developed an expert system for information on pharmacology and drug interactions (Roach et al. 1985). Under the supervision of John Roach, I picked the brains of pharmacologists Jeff Wilcke and Marion Ehrich during the knowledge acquisition phase of the knowledge engineering process. It was an eye opener for a computer science student who had to study pharmacology in order to create a working expert system. For knowledge representation, I organized and encoded the pharmacological information in rules and frames for systematic retrieval, including:

a. delineation, definition, and hierarchical subdivision of mechanisms responsible for drug interactions;
b. division of pharmacological agents into a hierarchy of subclasses to allow for defining interacting drugs by classes as well as by specific agents; and
c. correlation of drug classes and specific drugs with mechanisms by which they may be involved in drug interactions.

This information, accessible through a natural language-like and menu driven interface, allows clinicians to know:

a. what may happen when two drugs are used together, and why;
b. what can be done to alleviate detrimental interactions; and
c. what related drugs may also be involved in similar interactions.

A.I. and expert systems were mostly confined to the academic world and research communities until May 1997 when the IBM Deep Blue computer beat the world chess champion Garry Kasparov after a six-game match, marking the first time in history that a computer had ever defeated a world champion in a match play (IBM 2012).

A.I. made headline news again when it hit the spotlight on primetime television over three nights in February 2011: The IBM Watson computer won on "Jeopardy!" against two human champions and took home a $ 1 million prize (Paul 2011). Watson, named after IBM founder Thomas J. Watson, has the ability of encyclopedic recall and natural language understanding.

4   Gödel, Escher, Bach: Consciousness And Intelligence

Notwithstanding IBM's Watson or Apple's iPhone assistant Siri, the Pulitzer Prize-winning author of *Gödel, Escher, Bach: An Eternal Golden Braid* (GEB) would argue that Watson and Siri have very little to do with intelligence (Somers 2013). By exploring common themes in the lives and works of logician Kurt Gödel, artist M. C. Escher, and composer Johann Sebastian Bach, GEB author Douglas Hofstadter expounds concepts fundamental to mathematics, symmetry, and intelligence.

Hofstadter told *Wired* magazine's executive editor Kevin Kelly in a 1995 interview: "What Gödel, Escher, Bach was really about—and I thought I said it over and over again—was the word I. Consciousness. It was about how thinking emerges from well-hidden mechanisms, way down, that we hardly understand. How not just thinking, but our sense of self and our awareness of consciousness, sets us apart from other complicated things. How understanding self-reference could help explain consciousness so that someday we might recognize it inside very complicated structures such as computing machinery. I was trying to understand what makes for a self, and what makes for a soul. What makes consciousness come out of mere electrons coursing through wires" (Kelly 1995).

Is consciousness a prerequisite for intelligence or vice versa? Some researchers at the 2011 Brains, Minds and Machines Symposium suggested that a conscious system could be built from highly integrated intelligent systems with many nested information relationships (MIT150 Symposia: Brains, Minds and Machines Symposium 2011).

In his book, Hofstadter explains that "a 'program' which could produce brilliant music would have to wander around the world on its own, fighting its way through the maze of life and feeling every moment of it. It would have to understand the joy and loneliness of a chilly night wind, the longing for a cherished hand, the inaccessibility of a distant town, the heartbreak and regeneration after a human death. It would have to have known resignation and world-weariness, grief and despair" (Hofstadter 1999).

However, an A.I. 'program' can be programmed to believe that it has gone through all the hardship and valuable life experiences. In fact, some philosophers argue that human beings are artificial intelligences trapped in a fake universe; and some physicists are studying new scientific methods to either prove or disprove the theory that we are all part of a giant simulation created by higher beings or sentient machines (Merali 2013). Could there be some truth to the 1999 science fiction film *The Matrix?*

5 Hal and Star Trek's Holodeck: A.I., Arts, and Sciences

"People ask me if this is HAL," said David Ferrucci, lead developer of IBM's Watson, referring to the Heuristically programmed ALgorithmic (HAL) computer in *2001: A Space Odyssey* by Stanley Kubrick and Arthur C. Clarke. "HAL's not the focus; the focus is on the computer on *Star Trek*, where you have this intelligent information seeking dialogue, where you can ask follow-up questions and the computer can look at all the evidence and tries to ask follow-up questions. That's very cool"(Markoff 2011).

Google's cofounder Sergey Brin once said that he wanted to build a benign version of the sentient computer HAL (Hof 2013). Google's director of engineering and A.I. guru Ray Kurzweil wanted to build a machine smart enough to pass the Turing Test (Ingrahman 2013). To that end, Google's Knowledge Graph has catalogued over 700 million topics, locations, people, and other concepts; and created billions of relationships among them. Google's technical lead Shashi Thakur said, "Take Leonardo da Vinci. Let's say you wanted to learn about the most important Renaissance painters. You might search for 'Leonardo da Vinci' because he's the only Renaissance painter you know about. Now, you'll see information right in the search results that helps you explore the broader topic of Renaissance painters. You'll see some of the most famous painting from that era, like the Mona Lisa, and discover other painters of that time, like Michelangelo and Raphael" (Google 2013).

The ultimate *Star Trek* computer is a "holodeck" where true-to-life simulations can be run to seek information, assist humans with their decision-making, or simply take someone on a vacation. The first use of a holodeck by that name in the *Star Trek* universe was in the pilot episode of *Star Trek: The Next Generation*, "Encounter at Farpoint, Part I." A holodeck is a "smart" virtual reality system that combines transporter, replicator, and holographic systems (CBS Entertainment 2013). The A.I. programs, projected via emitters within a specially outfitted but otherwise empty room, can create both "solid" props and characters as well as holographic background to evoke any vista, any scenario, and any personality—all based on whatever real or fictional parameters are programmed.

Technologies involving artificial intelligence, arts, and sciences are taking steady steps towards the actualization of a holodeck:

1. In 1979, artist Harold Cohen created AARON, a computer program that produce art autonomously (Cohen 1979). In 2013, the three-piece robot band "Z-Machines" debuted in Tokyo, Japan (The Japan Time 2013). Created by engineers at the University of Tokyo, the band features keyboardist "Cosmo", guitarist "Mach" with 78 fingers, and drummer "Ashura" with 22 arms (McKenzie 2014). In Chap. 2, Gavin Sade describes experimental creative practices that bring together arts, science, and technology in imaginative ways. In Chap. 3, Mine Özkar expounds visual computation for good designs based on repetition and variation.

2. In August 2011, IBM unveiled its first Synapse chips featuring components that serve as 256 neurons and 262,144 synapses (Meyer 2011). In March 2012,

Hughes Research Laboratories (HRL) and the University of Michigan created a chip "memristor" that learns like the human brain by altering the synapses that connect its neurons (HRL Laboratories, LLC 2012). In April 2013, President Barack Obama announced $ 100 million in funding for arguably the most ambitious neuroscience initiative ever proposed: The Brain Research through Advancing Innovative Neurotechnologies, or BRAIN (Office of the Press Secretary 2013). The National Institutes of Health (NIH), Defense Advanced Research Projects Agency (DARPA), and National Science Foundation (NSF) are launching the BRAIN effort with funding in the President's FY 2014 budget. In Chap. 4, Raffaella Folgieri, Claudio Lucchiari, Marco Granato, and Daniele Grechi introduce BrainArt, a brain-computer interface (BCI) that allows users to create drawings using their own cerebral rhythms.

3. Researchers in "full space projection" employ neurobiology, story-telling, screenplay, visual effects, image stitching, and projection technology to create 360° immersive worlds (Overschmidt and Schröder 2013). The CAVE2 at the University of Illinois at Chicago is a life-size replica of the main bridge of the starship *U.S.S. Enterprise* (The Telegraph 2013). Google's Holodeck surrounds the user with moving pictures—an immersive way to experience locations and for Street View engineers to evaluate picture quality (Sullivan 2009). In Chap. 5, Nathan Cohen explores artificially created spaces that enhance spatial awareness and challenge our perception of what we encounter.

4. Microsoft's IllumiRoom is a proof-of-concept system that combines the virtual and physical worlds by using a Kinect for Windows camera and a projector to blur the lines between on-screen content and the real surrounding environment (Jones et al. 2013). In the ACM CHI 2013 Best Paper, Brett R. Jones et al. wrote that "peripheral projected illusions can change the appearance of the room, induce apparent motion, extend the field of view, and enable entirely new physical gaming experiences" (Jones et al. 2013). In Chap. 6, Keith Armstrong examines embodied experiences that affect the mind and body of participating audiences.

5. Regina Dugan, senior executive at Google and former director of DARPA, revealed in the 2013 D11 conference some of the most advanced wearable computing technologies: (a) an electronic tattoo on her left arm that can be used to authenticate a user in lieu of password, (b) a pill that can be ingested and then battery-powered with stomach acid to produce an 18-bit internal signal (Gannes 2013). In Chap. 7, Diomidis Spinellis uses a scientific experiment exhibit to teach the concept of physical computing.

6. Video games are becoming more realistic with advanced hardware and software. "Motion capture" and "performance capture" together record actors' movements, faces, and voice all at once. The result is a film-like quality to the action in the games (Frum 2013). In Chap. 8, Benjamin Cowley explains the massively multiplayer online game "Green My Place" aimed at achieving behavior transformation in energy awareness.

7. MakerBot Digitizer is a desktop device that scans almost any small 3-D object up to about 8 inches in diameter, and an identical (or optionally modified) object

can be manufactured right away by feeding the resulting file to a 3-D printer (MakerBot Industries, LLC 2013). For the 2013 Victoria's Secret fashion show, a 3-D scanner scanned the body of model Lindsay Ellingson and a 3-D printer printed an elaborate pair of "angel wings" costume (Segall 2013). It may come as a surprise that many of the M. C. Escher's optical illusion artwork have been realized as 3-D printed physical objects, notably by Prof. Gershon Elber at Technion, Israel Institute of Technology (Elber 2013). In Chap. 9, Robert Niewiadomski and Dennis Anderson portray 3-D manufacturing as the beginning of common creativity revolution and new applications such as bio printing, chemical engineering, and drug manufacturing. In Chap. 10, Stephen Barrass takes 3-D printing to another dimension by fabricating an object from a sound recording.

8. In 2010, Pindar Van Arman introduced Vangobot, a painting robot that makes what look to be impressionistic paintings (Van Arman 2010). In 2014, a team of physicists has demonstrated a device that can teleport quantum information to a solid-state quantum memory over a 25-kilometer optical fiber (Bussieres et al. 2014). In Chap. 11, Mari Velonaki examines the element of surprise and touch sensing in human-robot interaction; and in Chap. 12, Roman Danylak surveys the media machines in light of Marshall McLuhan's dictum "the medium is the message."

To quote computer graphics pioneer Ivan Sutherland who said in 1965: "The ultimate display would, of course, be a room within which the computer can control the existence of matter. A chair displayed in such a room would be good enough to sit in. Handcuffs displayed in such a room would be confining, and a bullet displayed in such a room would be fatal" (Sutherland 1965).

Sutherland's futuristic vision sounds just like *Star Trek's* holodeck!

6   Cyborg in the Arts and Sciences

Taking another step further or alongside *Star Trek's* holodeck is the idea of cyborg (short for cybernetic organism) coined in 1960 by Manfred E. Clynes and Nathan S. Kline in their article "Cyborgs and Space" about creating self-regulating man-machine systems to meet the requirements of extraterrestrial environments (Clynes and Kline 1960).

Some artists have tried to create public awareness of cybernetic organisms, ranging from paintings to installations. In November 2010, New York University arts professor Wafaa Bilal had a digital camera surgically implanted into the back of his head. As part of "The 3rd I" project commissioned by a museum in Qatar, the camera captured his everyday activities at 1-minute intervals 24 hours a day and streamed live on the Internet and at the Mathaf: Arab Museum of Modern Art (Ilnytzky 2010). Stelarc is a performance artist whose works have focused heavily on extending the capabilities of the human body using medical instruments, prosthetics, robotics, virtual reality systems, and other technologies. The following section presents an in-depth interview with Stelarc by Melbourne University Prof. Darren Tofts on January 16, 2008 in Australia (Tofts 2008).

Defense Advanced Research Projects Agency (DARPA) has supported research on implanting tiny Micro-Electro-Mechanical Systems (MEMS) devices into insect bodies while the insects are in their pupal stage (The Washington Times 2006). In 2006, Cornell University researchers have successfully created the first insect cyborgs—moths with integrated electronics in their thorax (Bozkurt et al. 2007). In 2013, a U.S. company backed by the National Institute of Mental Health sells a $ 100 education kit that lets anyone create a RoboRoach from a live cockroach and wirelessly control the left/right movement of the cockroach by microstimulation of its antenna nerves (Hamilton 2013).

In medicine, cyborgs can be restorative or enhanced. The former restores lost function, organs, or limbs whereas the latter exceeds normal human capabilities as dramatized in the popular TV shows *The Six Million Dollar Man* and *The Bionic Woman*. A brain-computer interface, or BCI, provides a direct path of communication from the brain to an external device, effectively creating a cyborg. In 2014, Georgia Tech professor Gil Weinberg created a prosthesis for amputee drummer Jason Barnes. By flexing his muscles, Barnes can send signals to a computer to tighten or loosen his grip on drumstick and control the rebound (Newman 2014).

7 An Interview With Cyborg Artist Stelarc (By Darren Tofts)

The Cyborg Cometh: It begins with a sinister laugh. Beatific, knowing and bordering on rapture, it revels in the possibilities of imminence, of something significant and profound about to be revealed. While portentous of the unknown, it is at the same time a very familiar, very human signature. But of what? For theorists such as Donna Haraway, Katherine Hayles, and Gilles Deleuze, it speaks of a new conception of the human, evolving with informatic technology into a hybrid biomachine. For the host of this sublime laughter, Stelarc, it speaks on behalf of his performative alter-ego, the body.

For more than 3 decades, Stelarc has explored the increasingly malleable relations between the body and technology. A pioneering exponent of cybernetic art, his work has questioned and broken down the technical and philosophical boundaries between human life as we know it and what it might become. This constant becoming-cyborg in Stelarc's work has transcended the built environment as well as the distributed spaces of the Internet and other telematic interfaces. Eager to identify unexplored somatic possibilities within and between the two, Stelarc is on the lookout for new performance stages, virtual, conceptual, and biological, where conditions of embodiment can be enacted and explored.

Darren Tofts Your collaborative 2005 work with Nina Sellars, *Blender*, was in some ways a departure for you, a shift from the immateriality of the virtual to the

Fig. 1.2 Blender (Tek-
nokunst, Melbourne). (Photo
by Stelarc. Courtesy of
Stelarc and Nina Sellars)

materiality of the physical, dealing with the visceral nature of bodily matter. Can you tell us how you see this work fitting in with your broader notion of the body?

Stelarc I guess the way that I've worked as an artist is not based on any particular medium and making any distinction between the virtual and the actual. In other words, all of these different performative modes and all of these projects involving alternate media have been expressing a certain conceptual concern of exploring the prosthetic augmentation of the body on one arm; of experimenting with alternate anatomical architecture on the other; and seeing the body not as necessarily a bio-logical body, but seeing it in the broader sense. There's also been a strategy that may be exemplified in Jean-Luc Nancy's postmodern approach in that his decon-struction's not through appropriation and juxtaposition, but rather deconstruction through exhausting a particular concept or idea. So the series of suspension perfor-mances was a means of exhausting the body into exposing its obsoleteness, and I like the idea that there is no strategy of deconstruction that involves appropriation, but rather the deconstruction occurs through the exhausting of an artistic act to expose its inadequacy and possibly reveal some alternate possibilities.

Blender (see Fig. 1.2) in a way is a counterpoint to the Stomach Sculpture (see Fig. 1.3) performance and project where initially I designed a sculpture for the

Fig. 1.3 Stomach Sculpture
(Fifth Australian Sculpture
Triennale, NGV, Melbourne).
(Photo by Tony Figallo.
Courtesy of Stelarc)

inside of my body, and in that case this was a sculpture that was inserted 40 cm into the stomach cavity. The object opened and closed, extended and retracted, had a flashing light and a beeping sound, so it was this kind of mechanic choreography within this soft and wet, vulnerable environment of the inside of the stomach. So blender could be seen as a counterpoint to that instead of a machine inhabiting the soft interior of the body. In *Blender*, a machine becomes a host for a liquid body, a liquid body composed of biomaterial from two artists.

Tofts Further to this, your initial collaborations with Oron Catts and Ionat Zurr on the *Extra Ear* project, as well as your ongoing development of this work is also very corporeal. For an artist who has for so long sought to escape the flesh, what attracts you to three ears rather than two?

Stelarc I've never really actually tried to escape the flesh as such, in effect all of these projects and performances were explorations in the psychological and physiological perimeters of the body. So even if you think of an intelligent agent as being embodied and embedded in the world, one can't really perform without a body. The question is what kind of a body, and do we simply accept the biological status quo of our present evolutionary trajectory, or do we consider ways of redesigning the body, ways of alternate mechanical augmentation of the body? Why not have an additional ear on the arm, maybe as a listening and transmitting device, an ear on the arm might be better position?

The Extra Ear project (see Fig. 1.4) actually begins way back in 1996. I was in Carnegie Mellon University and it was at that time when I grew rat muscle cells, but at that time I didn't really know what to do with them, didn't like the idea of putting them in a Petri dish on a pedestal, so the idea of growing something using living cells was something that was always a direction that I was intrigued with. The extra ear though, which was initially visualized in 1997 as an ear on the side of my head, kind of mimicked *The Third Hand* project (see Fig. 1.5) in the sense that the third hand was positioned beside my real right hand, and created this visual rhythm through repetition, so the idea of three ears was the continuity of the idea of a prosthesis in excess. But after years of trying to find surgical assistance and also

Fig. 1.4 Ear on Arm
(London, Los Angeles,
Melbourne). (Photo by Nina
Sellars; Courtesy of Stelarc)

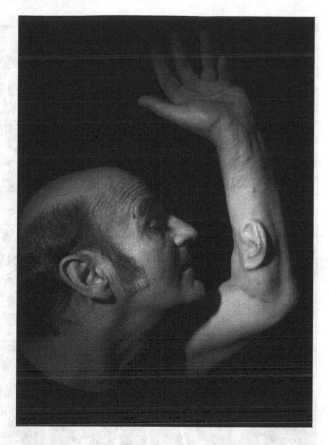

discovering that the ear on the side of the face was not an anatomically appropriate sight; it was near the jawbone, it was near facial nerves, which meant partial paralysis if an operation had gone wrong. In 2006 after years of additional attempt to get surgical assistance, I finally got funding and with the help of three surgeons in Los Angeles, we begun constructing an ear on my arm.

This is still only a relief of an ear, but it's the result of two surgical procedures: The first was to insert a skin expander, so over a period of a couple of months, I was injecting sterile saline solution, stretching the skin in that area, creating a pocket of excess skin that could be used in the surgical construction of the ear. That was removed in the second procedure and a med-pore scaffold was inserted, resulting in this present shape. Med-pore is a biomaterial that's used in reconstruction surgery; and it's a porous material, so it allows cell grows into it, so after a period of 6 months, tissue ingrowth and vascularization have occurred, so the ear is literally fused and fixed on my arm. We still need to lift the helix of the ear, creating a conch, and also highlighting the tragus area of the ear. Also, growing a soft ear lobe using my stem cells, so these are adipose-derived adult stem cells and using growth factors they can be directed to grow, for example, cartilage-like material so

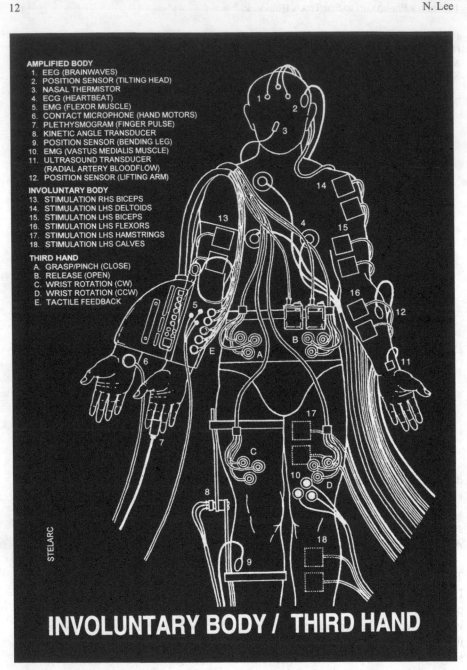

Fig. 1.5 The Third Hand Project (Yokohama, Melbourne 1990). (Diagram by Stelarc; Courtesy of Stelarc)

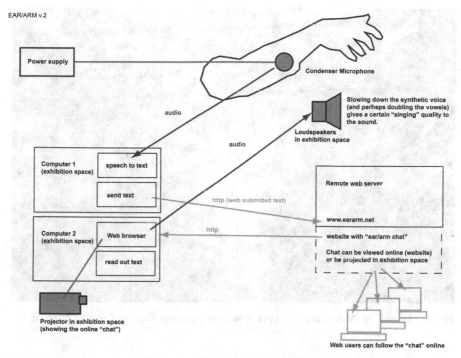

Fig. 1.6 Internet Ear (Aarlborg, Paris, Moscow). (Diagram by Mogens Jacobsen; Courtesy of Stelarc)

we'll be growing the earlobe. So it's partly a surgical construction—it involves tissue ingrowth, but also, in terms of more cutting edge research, it involves actually growing a part of the ear using adult stem cells. Finally, to implant a small microphone to connect to a bluetooth transmitter will enable the ear in any WiFi spot to be wirelessly connected to the Internet, so this effectively becomes a kind of Internet organ that's publicly accessible to people in other places (see Fig. 1.6). So if you're in Melbourne, you can listen to what my ear is hearing in London, or in New York, or wherever you are and wherever the artist is.

Tofts As with your famous retiring of the third hand a number of years ago, do you foresee a time when your extra ear will also be removed?

Stelarc With all of these performances, the body acts with a kind of indifference as opposed to expectation. When you do things with expectation, things quickly become predictable, possibilities quickly collapse into actuality. Performing with indifference, you allow the performance to unfold; you try to suspend that quick collapse into actuality, so the idea of performing with indifference is an important one. I've always been surprised during the performance with some of the performative things that have happened, and I've always been pleased by the unexpected ideas that have been generated by the various projects and performances. If this didn't happen, it wouldn't really be an activity that I'd continue to do, specifically in the series of performances that use Laser Eyes (see Fig. 1.7).

Fig. 1.7 Laser Eyes (Maki Gallery, Tokyo). (Photo by Takatoshi Shinoda; Courtesy of Stelarc)

Laser beams were projected initially from little mirrors on my eyes, but then guided through optic fiber cable to project from the eyes. I discovered by blinking and controlling the musculature around the eyes, I could literally draw with the laser beams in the space that I was performing in. So, in those actions, the eyes become not passive receptors of light and images, but rather active generators of images. And that was kind of a surprising outcome of initially just an idea of just probing inside the body; you could project from the body into the external space of the performance. But being able to draw with my two eyes and actually be able to draw and scribble individually with each eye—separate images—was surprising.

And also, the way that the third hand became one in which its actions were much more intuitively generated. Initially when I wanted my third hand to move, I'd have to think of the muscle I'd be using to actuate the motor, and I'd look at my third hand and it was a very sequential operation that lasted a second or two. But the more I used my third hand, the more intuitive it became, and the hand was almost moving of its own accord, and that was a really pleasing place to be when that was happening.

Tofts Similarly, other recent work, such as *Prosthetic Head* and *Walking Head* have revived the historical figure of the zombie as a means of extending your notions of obsolete body and alternate interfaces. What interests you about the zombie as a philosophical and technological concept and how do see these works operating as a kind of zombie aesthetic?

Stelarc Zombies and cyborgs are alternate ways to think about how the body functions anyway, we primarily function automatically, involuntarily. The notion of

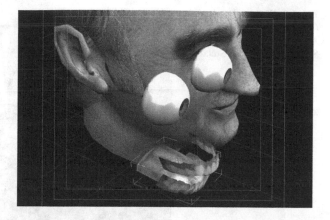

choice and free agency is limited within a certain frame of reference, so I think the prosthetic head and the walking head explore notions of artificial intelligence and artificial life in simple but performative ways. So, for example, the prosthetic head (see Fig. 1.8) is an agent that speaks to the person who interrogates it. Now, as an installation, it's a 5 m high projection, it has a sensor system that alerts it when someone's in the room, so when you walk into the space and to the keyboard, the head turns, opens its eyes, and asks you the first question. It has a database and real-time lip-syncing, so when you ask the prosthetic head a question, it scans the database, selects an appropriate response, and lip-syncs the answer in real time. So effectively what you've got is a talking head, but it's a computer generated head; it's simply a 3000 polygon mesh with skin on it. It's empty, there's nothing inside the head, but it has a performative and conversational behavior that coupled to a human body and capable of generating some verbal and visual exchanges, which is why I call it a prosthetic head.

The Walking Head (see Fig. 1.9) is a more chimeric construct: an insect-like six-legged walking machine that has a computer generated human-like head mounted on the chassis of the machine. It is an actual virtual interface in that the facial behavior of the walking head is actuated and modulated by the mechanical movement of the legs. So the robot sits and waits until someone approaches it. Its ultrasound sensor system detects that you're in front of it; the robot then stands, selects from its library of possible movements, and then performs a simply choreography of a few minutes before it sits down and waits for the next person to come along. One shouldn't see these as simulating artificial intelligence or artificial life, but rather one should see it as alternate bodily constructs and actual virtual interfaces that allow the body to become an extended operational system.

Tofts In 2007, you explored Second Life for the first time as an interactive, virtual world. Can you tell us about that?

Stelarc Second Life is an alternate motive of interaction and operation on the net, so as a performative side, it's got lots of interesting possibilities (see Fig. 1.10). I think we really have to recode the aesthetic and interactive possibilities so we're not just simply mimicking the visual and functional appearances of the real world. My

Fig. 1.9 Walking Head
(Heide Museum of Modern
Art, Melbourne). (Photo by
Stelarc; Courtesy of Stelarc)

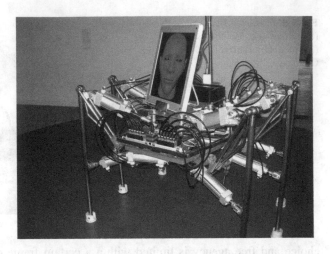

approach would be to flush out Second Life a bit more. Second Life for me is a kind of second skin, in other words the avatar becomes an alternate mode of interface with people in other places, as well as providing an appropriate Internet pulse to it; in other words, perhaps your avatar can become a barometer of Internet activity in a real-time sense. So there are just some thoughts about what is seductive for me about the notion of Second Life.

Tofts What do you think Second Life has to offer as a potential performance space? And as a second part to the question, what would you like to do in Second Life?

Stelarc I've always thought in working with machines and computers that you're giving the body an extended operational capability. In other words, it's never been an issue to me as in what's in control; it's not that the body has a master control of the computer, but rather, plugging the body into these technological systems, what alternates and surprising modes of operation are possible. This includes the idea of the body becoming a split physiology; in other words, if people in other places are prompting the left side of my body, and the left side of my body is responding involuntarily to their remote choreography; but simultaneously I can counterpoint by actuating a third hand on the right side of my body, so there's voltage in on one side, involuntarily generating movements that I can counterpoint voltage out on the right side of my body (see Fig. 1.11). Then, this idea of the body is not as a side of a single agency but effectively a host for multiple and remote agents. That generates interesting performative possibilities; so using machines and computers is really about the idea of erasing simplistic notions of single agency, of simplistic identity, and thinking more of the body in fragments or the split physiology, being remotely prompted. That's what's interesting about using technology.

Tofts What do you see as the possible future of human-computer collaboration for the creation of art and for artistic expression?

Fig. 1.10 Second Life Laser Eyes (Second Life Performance). (Screen capture by Stelarc; Courtesy of Stelarc)

Fig. 1.11 Split Body (Galeria Kapelica, Ljubljana). (Photo by Igor Andjelic; Courtesy of Stelarc)

Stelarc One possible scenario amongst other cyborg constructs that I think goes beyond the military notions of traumatized bodies with replacement parts or massive exoskeletal augmentations of the body is this idea that perhaps all technology in the future would be invisible because it'd be inside the body. In other words, as technology becomes micro-miniaturized, it can be incorporated inside the body. Given we're now constructing devices on a nanoscale, then it's conceivable that we can re-colonize the human body, augmenting our bacterial and viral populations, producing a better internal surveillance of the body. I think the problem with surveillance is not so much whether ethnically we should have more or less external surveillance, but rather the imperative of having more internal surveillance because we will be dealing not only with internal threats but cellular spaces, and that's the implication of working at a nanoscale level. It's quite conceivable that we can construct sensors that can detect pathological changes in chemistry and in temperature. We'd certainly be able to detect blockage in our circulatory system and not only develop the sensors, but micro-miniaturized and nanoscale robots that could be either automatically maintaining the inside of the body or remotely guided by an external surgeon, so this is quite conceivable. So what's interesting artistically and aesthetically about working at a nanoscale is the different kind of movements and mechanisms and structures that you could sculpturally and artistically employ. I think that's what's interesting.

8 Epilogue

In a 1996 interview by Terry Gross, Steve Jobs said that "science and computer science is a liberal art; it's something everyone should know how to use, at least, and harness in their life. It's not something that should be relegated to 5 % of the population over in the corner. It's something that everybody should be exposed to and everyone should have mastery of to some extent, and that's how we viewed computation and these computation devices" (Gross 2011).

Since the world's first computer art was created in the form of a pin-up girl in late 1950's on a $ 238 million military computer, today's digital artists can unleash their creativity and imagination using ubiquitous software tools such as Illustrator, Photoshop, and Corel Painter on an affordable home computer. As Albert Einstein once said, "True art is characterized by an irresistible urge in the creative artist." The appendices in this book showcase some of the artwork by young contemporary digital artists Aleksei Kostyuk, Pawel Nolbert, and Tanya Varga.

Acknowledgements I would like to acknowledge my colleague Maria Rizzo for introducing me to Darren Tofts who interviewed Stelarc; my intern Joey Lee (Ngee Ann Polytechnic) for his assistance in transcribing the Stelarc interview video; and the National University of Singapore and Media Development Authority for their wonderful internship program.

References

Aron, Jacob. Software tricks people into thinking it is human. New Scientist. [Online] September 6, 2011. http://www.newscientist.com/article/dn20865-software-tricks-people-ino-thinking-it-is-human.html.

Association for Computing Machinery. A.M. Turing Award. Association for Computing Machinery. [Online] 2012.

Association for Computing Machinery. Edward H Shortliffe. Grace Murray Hopper Award. [Online] [Cited: November 12, 2013.] http://awards.acm.org/award_winners/shortliffe_1337336. cfm.

Bozkurt, A., et al. Microprobe microsystem platform inserted during early metamorphosis to actuate insect flight muscle. IEEE Xplore. [Online] January 21-25, 2007. http://ieeexplore.ieee. org/xpl/articleDetails.jsp?arnumber=4432976.

Burton, Robert. Ivan Sutherland. A.M. Turing Award. [Online] Association for Computing Machinery, 1988. http://amturing.acm.org/award_winners/sutherland_3467412.cfm.

Bussieres, Felix, et al. Quantum teleportation from a telecom-wavelength photon to a solid-state quantum memory. Cornell University Library. [Online] January 27, 2014. http://arxiv.org/ abs/1401.6958.

CBS Entertainment. Holodeck. Star Trek. [Online] [Cited: November 16, 2013.] http://www. startrek.com/database_article/holodeck.

Clynes, Manfred E. and Kline, Nathan S. Cyborgs and Space (Reprinted with permission from Astronautics, September 1960). New York Times. [Online] 1997. http://partners.nytimes.com/ library/cyber/surf/022697surf-cyborg.html.

Cohen, Harold. What is an Image. Harold Cohen Online Publications. [Online] 1979. http://www. aaronshome.com/aaron/publications/index.html.

Copeland, B.J. DENDRAL. Encyclopaedia Britannica. [Online] [Cited: November 12, 2013.] http://www.britannica.com/EBchecked/topic/745533/DENDRAL.

Edwards, Benj. The Never-Before-Told Story of the World's First Computer Art (It's a Sexy Dame). The Atlantic. [Online] January 24, 2013. http://www.theatlantic.com/technology/ archive/2013/01/the-never-before-told-story-of-the-worlds-first-computer-art-its-a-sexy-dame/267439/.

Elber, Gershon. Escher for Real. Technion. [Online] [Cited: November 30, 2013.] http://www. cs.technion.ac.il/~gershon/EscherForReal/.

Frum, Larry. Emerging technology heightens video-game realism. CNN. [Online] November 14, 2013. http://www.cnn.com/2013/11/14/tech/gaming-gadgets/realism-video-games/index.html.

Gannes, Liz. Electronic Tattoos and Passwords You Can Swallow: Google's Regina Dugan Is a Badass. All Things D. [Online] May 29, 2013. http://allthingsd.com/20130529/electronic-tattoos-and-passwords-you-can-swallow-googles-regina-dugan-is-a-badass/.

Google. The Knowledge Graph. Google. [Online] [Cited: December 26, 2013.] http://www.google. com/insidesearch/features/search/knowledge.html.

Gross, Terry. Steve Jobs: 'Computer Science Is A Liberal Art'. NPR. [Online] October 6, 2011. http://www.npr.org/2011/10/06/141115121/steve-jobs-computer-science-is-a-liberal-art.

Hamilton, Anita. Resistance is Futile: PETA Attempts to Halt the Sale of Remote-Controlled Cyborg Cockroaches. Time Magazine. [Online] November 1, 2013. http://newsfeed.time.com/ 2013/11/01/cyborg-cockroaches-are-coming-but-not-if-peta-has-anything-to-say-about-it/.

Hof, Robert D. Deep Learning. MIT Technology Review. [Online] April 23, 2013. http://www. technologyreview.com/featuredstory/513696/deep-learning/.

Hofstadter, Douglas. Gödel, Escher, Bach: An Eternal Golden Braid. Google Books. [Online] February 5, 1999. http://books.google.com/books?id=aFcsnUEewLkC.

HRL Laboratories, LLC. Artificial synapses could lead to advanced computer memory and machines that mimic biological brains. HRL Laboratories, LLC. [Online] March 23, 2012. http:// www.hrl.com/hrlDocs/pressreleases/2012/prsRls_120323.html.

IBM. Deep Blue. [Online] IBM. [Cited: November 5, 2012.] http://researchweb.watson.ibm.com/deepblue/.

Ingrahman, Nathan. How Larry Page and the Knowledge Graph helped Ray Kurzweil decide to join Google. The Verge. [Online] March 20, 2013. http://www.theverge.com/2013/3/20/4127266/ray-kurzweil-recruited-by-larry-page.

Jones, Brett R, et al. IllumiRoom: Peripheral Projected Illusions for Interactive Experiences. Microsoft Research. [Online] 2013. http://research.microsoft.com/en-us/projects/illumiroom/.

Jones, Brett R, et al. IllumiRoom: Peripheral Projected Illusions for Interactive Experiences. Microsoft Research. [Online] 2013. 1. http://research.microsoft.com/en-us/projects/illumiroom/IllumiRoom_CHI2013_BJones.pdf.

Kelly, Kevin. By Analogy. Wired. [Online] November 1995. http://www.wired.com/wired/archive/3.11/kelly.html.

Ilnytzky, Ula. Wafaa Bilal, NYU Artist, Gets Camera Implanted In Head. The Huffington Post. [Online] November 23, 2010. http://www.huffingtonpost.com/2010/11/23/wafaa-bilal-nyu-artist-ge_n_787446.html.

MakerBot Industries, LLC. MakerBot Digitizer. MakerBot Industries, LLC. [Online] [Cited: November 16, 2013.] http://store.makerbot.com/digitizer.html.

Markoff, John. Computer Wins on 'Jeopardy!': Trivial, It's Not. [Online] The New York Times, February 16, 2011. http://www.nytimes.com/2011/02/17/science/17jeopardy-watson.html?pagewanted=all.

McKenzie, Sheena. Meet the robot guitarist with 78 fingers and coolest cable hair you've ever seen. CNN. [Online] March 14, 2014. http://www.cnn.com/2014/03/14/tech/meet-the-robot-guitarist/index.html.

Merali, Zeeya. Do We Live in the Matrix? Discover Magazine. [Online] November 15, 2013. http://discovermagazine.com/2013/dec/09-do-we-live-in-the-matrix#.UoU3_GR4ayg.

Meyer, David. IBM 'neuron' chips mimic brain processing. ZDNet. [Online] August 18, 2011. http://www.zdnet.com/ibm-neuron-chips-mimic-brain-processing-3040093720/.

MIT150 Symposia: Brains, Minds and Machines Symposium. Consciousness and Intelligence. MIT Tech TV. [Online] June 21, 2011. http://techtv.mit.edu/videos/13236-consciousness-and-intelligence.

Myers, Andrew. Stanford's John McCarthy, seminal figure of artificial intelligence, dies at 84. [Online] Stanford University, October 25, 2011. http://news.stanford.edu/news/2011/october/john-mccarthy-obit-102511.html.

Newman, Lily Hay. This Drummer Has a Third Arm. Slate. [Online] March 10, 2014. http://www.slate.com/blogs/future_tense/2014/03/10/robot_drumming_prosthesis_from_georgia_tech_gives_this_drummer_a_third_arm.html.

Office of the Press Secretary. Fact Sheet: BRAIN Initiative. The White House. [Online] April 2, 2013. http://www.whitehouse.gov/the-press-office/2013/04/02/fact-sheet-brain-initiative.

Overschmidt, Gordian, and Schröder, Ute (Hrsg.). Fullspace-Projektion. Springer. [Online] 2013. http://www.springer.com/computer/image+processing/book/978-3-642-24655-5.

Paul, Ian. IBM Watson Wins Jeopardy, Humans Rally Back. [Online] PCWorld, February 17, 2011. http://www.pcworld.com/article/219900/IBM_Watson_Wins_Jeopardy_Humans_Rally_Back.html.

Roach, John, et al. An expert system for information on pharmacology and drug interactions. Computers in Biology and Medicine. [Online] Volume 15 Issue 1, 1985.

Saenz, Aaron. Cleverbot Chat Engine Is Learning From The Internet To Talk Like A Human. SingularityHUB. [Online] January 13, 2010. http://singularityhub.com/2010/01/13/cleverbot-chat-engine-is-learning-from-the-internet-to-talk-like-a-human/.

Segall, Laurie. Victoria's Secret model wears 3-D printed wings. CNNMoney. [Online] December 10, 2013. http://money.cnn.com/2013/12/10/technology/victorias-secret-3d/index.html.

Somers, James. The Man Who Would Teach Machines to Think. The Altantic. [Online] October 23, 2013. http://www.theatlantic.com/magazine/archive/2013/11/the-man-who-would-teach-machines-to-think/309529/.

Sullivan, Danny. Google Holodeck: StreetView In 360 Degrees. Search Engine Land. [Online] May 27, 2009. http://searchengineland.com/google-holodeck-streetview-in-360-degrees-19808.

Sutherland, Ivan E. The Ultimate Display (1965). CiteSeer. [Online] Proceedings of the IFIP Congress, 1965. http://citeseer.ist.psu.edu/viewdoc/summary?doi=10.1.1.136.3720.

The Japan Time. Z-Machines debut show. YouTube. [Online] June 25, 2013. https://www.youtube.com/watch?v=GK9n97fN1Go.

The Telegraph. Star Trek-style holodeck becomes reality as scientists invent 3D vitual reality system. The Telegraph. [Online] February 20, 2013. http://www.telegraph.co.uk/science/science-video/9882646/Star-Trek-style-holodeck-becomes-reality-as-scientists-invent-3D-vitual-reality-system.html.

The Washington Times. Military seeks to develop 'insect cyborgs'. The Washington Times. [Online] March 13, 2006. http://www.washingtontimes.com/news/2006/mar/13/20060313-120147-9229r/.

Tufts, Darren. Interview with Stelarc. ACM Digital Library. [Online] October 2008. http://dl.acm.org/citation.cfm?doid=1394021.1394023.

Turing, A. M. Computing machinery and intelligence. Mind. [Online] 1950. http://loebner.net/Prizef/TuringArticle.html.

Van Arman, Pindar. Representational Painting Robot – with Impressionistic Slant. Art and Robots. [Online] November 12, 2010. http://www.zanelle.com/.

Weizenbaum, Joseph. ELIZA—a computer program for the study of natural language communication between man and machine. Communications of the ACM. [Online] January 1966. http://dl.acm.org/citation.cfm?id=365168&dl=ACM.

Young, Denise. The Edge of Possibility: Regina Dugan. [Online] Virginia Tech Magazine, Summer 2013. http://www.vtmag.vt.edu/sum13/regina-dugan.html.

Chapter 2
Experimental Creative Practices

Gavin Sade

1 Introduction

From the earliest human creative expressions there has been a relationship between art, technology and science. In Western history this relationship is often seen as drawing from the advances in both art and science that occurred during the Renaissance, and as captured in the polymath figure of da Vinci. The twentieth century development of computer technology, and the more recent emergence of creative practice-led research as a recognized methodology, has lead to a renewed appreciation of the relationship between art, science and technology.

This chapter focuses on transdisciplinary practices that bring together arts, science and technology in imaginative ways. Showing how such combinations have led to changes in both practice and forms of creative expression for artists and their partners across disciplines. The aim of this chapter is to sketch an outline of the types of transdisiplinary creative research projects that currently signify best practice in the field, which is done in reference to key literature and exemplars drawn from the Australian context.

2 Art + Science

In his work *Behind appearance: a study of the relations between painting and the natural sciences in this century* Waddington, a biologist writing about painting and natural sciences, suggests that "science is not merely a one-eyed Cyclops" but instead, humans have "innumerable eyes, all yielding their overlapping insights to his one being, that struggles to accept them in all their variety and richness" (1969). It is in this spirit that we set out to addresses the subject of *Digital da Vinci*—that is to say this chapter discusses creative practices that transcend traditional disciplinary boundaries in the same manner as Leonardo da Vinci—working across art, science and technology. To approach this subject we turn to the field of art-science, and consider this field from the perspective of literature on the philosophy of science,

G. Sade (✉)
School of Design, Queensland University of Technology, Brisbane, Australia
e-mail: g.sade@qut.edu.au

N. Lee (ed.), *Digital Da Vinci,* DOI 10.1007/978-1-4939-0965-0_2,
© Springer Science+Business Media New York 2014

experimental art, and interdisciplinarity. In this vein there is a focus on practices that do not replicate existing disciplinary forms, divisions of labor, or hierarchies of knowledge, but instead seek new synthesis and negotiate what Barry et. al. describe as "forms of agonism and antagonism that often characterize relations between disciplinary and interdisciplinary research" (2008).

Guattari describes such artistic cartographies as having "always been an essential element of the framework of every society" (Guattari 1995, p. 130). This is no different in science to any other domain of society, and from this perspective contemporary art-science projects that provide one of the more compelling responses to the question posed by *Digital da Vinci*—how can we encourage and empower a new generation of "well-rounded" scholars and students, through unconventional and creative application of computer science? This question will be addressed through an examination of several key Australian examples of art-science projects. The art-science community in Australia is particularly vibrant—considering the size of the country—and actively supported by the Australia Council for the Art, through the Experimental and Emerging Arts program.

Studies of inter and trans disciplinarity provide one way of approaching art-science. In this chapter we will reference two key bodies of work, that of Nowotny (2001) and that of Born and Barry (2010) who extend Nowotny's work in a way that allows a more nuanced evaluation of art-science projects. The chapter then turns to ask what makes experimental art experimental, and considers this question not from an art historical perspective, but through the lens of the philosophy of science, and specifically in relationship to the history of scientific experiement(ation). Thus reframing the concerns of artists about the intrumentalisation of art when undertaken within scientific paradigm, by articulating this in relationship to the contested relationship between experimenter, instrument/experimental apparatus, and theory see within the broader literature.

Despite questions about the nature of creative experimentation and whether it constitutes research[1], has been an increased recognition of Artistic Research[2] within Australia over the last 5 years. This has been a result of changes in Government research policy[3] that lead to the recognition of artistic and creative outputs as research outputs. When combined with the discussions on inter/trans disciplinarity, and experimentation, we see that experimental art-science projects engage in a what Willis (2006) describes as ontological design, This we suggest points toward a significant methodological development for Artistic Research more broadly. Similarly such

---

[1] This debate can be seen in the broader literature on Artistic Research, creative practice-led research, for example Working papers in Art and Design Research, Art&Research: A Journal of Ideas, Contexts and Methods. This concern is also echoed in literature focused on innovation and research & development. For example NESTA reports by Bakhshi and Throsby (2010), Bakhshi et al. (2011).

[2] In this chapter the term Artistic Research will be used to refer to creative practices undertaken in the context of research, which is drawn from the work of Coessesn et al. (2009).

[3] For example the Australian Excellence in Research policy that recognizes creative works as research output.

an ontological perspective suggests art-science projects may be exemplars of what Bakshi et al. (2010; 2011) describe as innovation through experiment. As a result this chapter aims to present one way of approaching of the epistemic tensions and transformative potentials of the larger *da Vinci* agenda.

3   Art and Science—Genealogy of Sorts

Art-Science, and Experimental Art, are not new areas of practice—and both can be positioned in respect to a range of historical threads (genealogies of practice). For the purposes of this chapter, and in line with the focus of *Digital da Vinci* on computer science, we will begin with the movements of computer art, cybernetic art and experimental art—which emerged in the 1960s as a result of the convergence of art and the nascent fields of Information and Communication Technology. These movements are relatively marginal; either being cited in respect to other named movements, or regarded simply as technological experimentation. Yet in the context of this chapter we consider artists working with new technologies mid last century, such as early computers, and exploring the implications of emerging scientific theories as precursors of today's art-science practice.

Over the last decade there has been a renewed interest in these movements, and a range of publications that map this terrain, for example Art of the Electronic Art (Popper 1993), Digital Arts (Paul 2003), New Media in Art (Rush and Rush 2005), Art in the Digital Age (Wands 2006), and Art and Electronic Media (Shanken 2009) to name but a few. Similarly there are a number of voices, such as Manovich (2001); Bourriaud (2002) and Quaranta (2010), who have examined the distinctions between contemporary art and New Media Art, and the emergence of a "post-media aesthetic" (Manovich 2001). More recently Bridles's (2011, 2012) *New Aesthetic*[4] has emerged as a "catch-all" for almost all forms of art and design that involve digital technologies and computation reshaping the aesthetic experience of everyday life—and increasingly life itself. While there has been substantial interest in the ways technology (and science) are changing creative practices and art, new artistic forms and practices related to science and technology have continually struggled to find a place within domains of science or art. This is not so much a result of some form of resistance from the any so called conservative establishment, but is really due to the difficulty in appreciating (valuing) new practices that cross over disciplinary boundaries, or result in new synthesis of disciplines.

Art science is one of those areas that poses such problems. For Born and Barry (2010) the practices and outcomes need to be understood, and valued, in respect to a broader context that encompasses multiple disciplinary perspective and histories. This they suggest includes: conceptual and post conceptual art; historical movement of art and technology; and the broader development of computation,

---

[4] As documented by Bridle at the following URL: http://new-aesthetic.tumblr.com/.

biological sciences and technologies, with an origin in theories of cybernetics. The challenge of understanding such practices is that they do not comfortably fit within a disciplinary framework, and are commonly found outside the normal sites for disciplinary practice. For example the computer artists of the 60s and 70s often found themselves working in newly formed interdisciplinary departments, which had begun to explore the use of computers within an art setting. However, these groups often found themselves outside the normal practices of existing disciplines. Brown (2008) describes one such interdisciplinary group, the Experimental and Computing Department within the Slade School of Fine Art, and the difficulty of artists working in this area in the early 70s to gain recognition within exiting disciplines. However, the work coming from these early interdisciplinary groups was instrumental in the development of the broader field of computer graphics[5], and has been recognized in retrospective exhibitions[6]. Today this type of practice is described as "blurring the boundaries between genres and disciplines, [as well as] redefining the contexts of use and modes of distribution" (Freyer et al. 2008, p. 10). This is often seen as a result of a continual focus on the "new"—as is suggested in names like New Media Art.

Artists working with new technologies have often been criticized for a form of technological fetishism (Manovich 2003), where the focus on new technology is in part viewed as over-determining artistic practices and outcomes. However, to engage in a constructive reading of such practices requires, as Paul (2008, p. 5) argues, more than a "strictly art-historical perspective." Understanding the significance of such practices requires an appreciation of the multiple disciplinary trajectories that converse within a specific instance of practice, which draws together fields of science, technology and media theory. This is not to disregard an art-historical perspective, for example authors such as Gere (2008); Popper (1993) and Shanken (2009) trace the linage of New Media Art practices to the Futurist, Surrealist, Dadist, Fluxus, Systems Art and Cybernetic Art movements of the last century, and note the influence of figures such as Dechamp, Nam June Paik, Cage and Sol LeWit. What emerges from the literature is an appreciation of the diversity of art-science. Art-science spans the full spectrum of both scientific endeavor and art practices—from critical engineering practices of Jerimijenko to the radical post-humanist work of Stelarc and the bio-arts of Kac, Zurr & Catts. These practices are all unique combinations of science and art, which in each configuration draw upon different histories of science and art.

---

[5] Computer Graphics is an examples of a field that has been the result of interdisciplinary research—from the early computer artists and researchers working together to develop the potential for new visual technology, to the contemporary animation studios working across Film, TV and Computer games pushing boundaries of the field through creative application.

[6] For example the computer art collection at the Victoria and Albert Museum London, see Beddard (2009).

Much of what is named art-science takes place within a research setting, as is seen in Wilson (2002) survey of what he calls *Information Arts*, in which he outlines several differing artistic approaches to engaging with techo-scientific research. This survey spans scientific disciplines, and concludes that artistic research within art-science takes of different forms, including: exploration of new possibilities opened up by science and technology; critical engagement and questioning of the cultural implications of specific lines of research; the use/application of new capabilities to address themes not directly related to specific science of technology; to the incidental use of technology within practices. (Wilson 2002, pp. 8–9) In the context of this chapter we are most interested in practices which do not simply "use" new technology, but have a critical or applied role in a specific line of scientific or technical research—rather than being a "distant commentator" or consumers of the outcomes of research without taking part in the processes of knowledge creation.

4   From Inter to Trans Disciplinary

These types of art-science project exhibit transdisciplinary characteristics, and to develop an appreciation for art-science requires an understanding of the movement from discipline to inter disciplinary to trans disciplinary. As Ox and Lowenberg point out "art-science, refers to worldviews, conceptual systems and research based on equal contribution from differently trained minds." (2013) In the first issue of *Leonardo*, a journal dedicated to writings about art science and technology inaugurated in 1968, Waddington writes that it would be a "mistake to see the traffic between art and science as one-way" (1968). Yet in outlining this interaction, art and science remained within fixed disciplinary boundaries influencing each other through their expressions. Around the same time C. P. Snow presented his two cultures argument, which marks a moment in recent history cited in much of the literature on interdisciplinarity.

> the clashing point of two subjects, two disciplines, two cultures—of two galaxies, so far as that goes—ought to produce creative chances. In the history of mental activity that has been where some of the breakthroughs came. (Snow 1964, p. 6)

Since C. P. Snow's 1964 two cultures argument, there has also been an increasing focus upon of interdisciplinary. More recently Csikszentmihalyi (1999, p. 314) argues that it is only within interdisciplinary settings where "individuals, domains and fields intersect" that the process of creativity can be observed, is particularly poignant. Carter (2004) suggests that this process of creativity is a form of *poiesis*, or place-making. This he describes as "collaborators plunge[ing] into the realm of Becoming" (Carter 2004, p. 11). The resulting tensions and exchanges bring into question the assumed "natural places of ideas, images and materials" (Carter 2004, p. 11). The outcomes of art-science collaborations are expressions of this negotiation, as ideas and materials as they are reorganized into creative forms and experiences. This is a shared creative act of placing things back together, a process that Carter argues, produces knowledge through the way collaborators combat the ideological character of their respective disciplinary discourses and myths by inventing artificial myths.[7] This suggests the importance of the arts and creative practice-led research within interdisciplinary collaborations, which Bennett describes as follows.

> What is clearer today than in previous generations of research is that the aesthetic (in the fullest sense, encompassing the practical study of affect, sensation, perception, behavior, imagination) is fundamental to any understanding of the connections between lifeworlds, disciplinary procedures and given problems: the arts, in other words, are at the core of the transdisciplinary experiment. (Bennett 2012)

The urgency of the so called "trans disciplinary experiment" is driven by the realization that the problems humanity faces, for example complex health issues or global climate change, are highly complex—even wicked (Horst 1973)—and require the combined efforts of multiple disciplines. As a result most of the work on interdisciplinarity has been focused on knowledge production, aka research, and can be seen across a range of disciplinary areas spanning art, design, social science, engineering, to medical sciences. Much of this writing reflects what Nowotny et al. (2001) describe as the movement from Mode-1 to Mode-2 knowledge production. Where Mode-1 knowledge production focuses on highly specialized disciplinary research, Mode 2 is carried out in respect to application, and involves heterogeneous teams of researchers and partners from multiple disciplines. Nowotny et. al use the prefix trans for Mode 2 knowledge. This form of knowledge production is described as "inherently transgressive" in that it "transcends disciplinary boundaries [...] reaching beyond interdisciplinary to trans disciplinary" (Nowotny et al. 2001, p 89)."

Within the literature there is a clear distinction made between multi, inter and trans. Multi and inter disciplinary services the "mutual needs of two disciplines", while transdisciplinary work "is impelled by external conditions or problems, but also by the conviction that disciplines do not have proprietary rights over their domains" (Bennett 2012). Across the literature there is a shared focus on complex, multi dimensional, highly relational, interdependent problems, which necessitate a methodological approach that transcends the singular foci of existent disciplines. Combined with the general view that the resulting synthesis cannot be reduced or evaluated from singular disciplinary perspectives; a "theoretical, conceptual, and methodological reorientation with respect to core concepts of the participating disciplines" (McMichael 2000)

Art-science is one of the key examples of this type of trandisciplinarity, and to develop an appreciation for the different types and forms of art-science we will employ the framework outlined in the *Logics of Interdisciplinarity* Barry et al. (2008).

---

[7] Here Carter is referencing Barthes' *Mythologies* (1973).

They describe the several logics and three modes that frame and organize practices. The most commonly described mode of practice encountered is referred to as the *integrative/synthesis mode* that is "conceived in terms of the integration of two or more 'antecedent disciplines' in relatively symmetrical form" (Barry et al. 2008). The second mode is referred to as the *subordinate/service mode*, where one discipline is in service to another. For example, a technology partner provides a service to an artist, filling a gap based on disciplinary expertise, or an artist is "employed" to visualize scientific data in order to present findings to the public. In both of these instances the partners effort (work) remains within their respective disciplines.

However, not all interdisciplinary collaborations fall into these two modes, a third modes involves a "commitment to contest or transcend the given epistemological and ontological assumptions of historical disciplines" (Barry et al. 2008). This is referred to as the *agonistic/antagonistic* mode, and the mode which most closely reflects descriptions of transdiciplinarity. Many art-science projects exhibit characteristics of this mode, especially where the artistic practices involves a critique, or questioning, of science, or where the working methods of science infect artistic processes and outcomes. Such projects are not easily reduced to the 'antecedent disciplines'. In other words, something new is born which requires a new frame of reference before it can be fully appreciated and evaluated. These three modes are not mutually exclusive, and many projects display characteristics of more than one—especially the third mode, as this type of critical questioning that is part of the artistic method—as seen in contemporary practices within the fields of conceptual and experimental art.

Similarly, interdisciplinary projects follow a series of different logics, which are often invoked when establishing the rationale and justifications for a project. Barry's study of interdisciplinary projects, including art-science, shows that many projects follow what they describe as the logics of accountability and innovation. For example there are numerous examples in the literature which position artists and creativity within innovation life cycles[8], similarly artists work is often seen as providing a form of public account of science (include citations). In many instances these logics become performative[9] and as a result structure practice—from criteria for funding schemes to evaluation methods. Thus despite the range of activity in art-science there remains a predominance of projects that are justified in relationship to either; their role in innovation; or as a way of representing science to a public. In contrast to these two logics, of innovation and accountability, there are some examples of art-science projects that follow a logic which Barry and Born name as the "logic of ontology".

---

[8] This is the basis of much of the work on Creative Industries and Innovation, see Bakhshi et al. (2010; 2011).

[9] Performative is used here in respect to concepts of performativity and practice, with origins in the work of Austin (1962). Similarly Pickering (2010) describes science as performative.

certain art-science initiatives are concerned less with making art or science accountable or innovative than with altering existing ways of thinking about the nature of art and science, as well as with transforming the relations between artists and scientists and their objects and publics. (Barry et al. 2008)

This focus on ontology and change is similar to notions of ontological design (Willis 2006) and Fry's (2009) redirective practice. This is an important concept in respect to art-science, and transdiciplinarity, as transdiciplinary projects very often involve a "redirection of the habitual, a change in the being of the practitioner" (Fry 2009, p. 20). This is a recognition of the ontological nature of transdisciplinary practices, and the ways practice is involved in self and world making. The work of Fry, a design theorist, shows that we are designed by, and design within, the designed world, and that our designs continue to design long after leaving the drawing board, studio or laboratory. In his studies of the practice of science Pickering presents a somewhat similar description of the material agency of the machines of science as a decentered posthumanism. Similarly Winograd and Flores (1986) discuss the ontological nature of design, showing how the design of a "new technology or systemic domain create new ways of being that previously did not exist and a framework for actions that would not have previously made sense" (Winograd and Flores1987, p. 177). Ontological design(ing), Willis (2006) claims, is both a "hermeneutics of design [...] understood as a subject-decentered practice" as well a case for mindful intervention within this circular operation of design, which in the context of the contemporary crisis of crisis is necessarily political. This recognition of the political brings us back to Barry et al. (2008)—who suggests that one mode for interdisciplinary projects has a ground in Mouffe's political theory of agonism (2013), a subject developed in respect to practices in art and design by DiSalvo (2012). Many of the artists and designers discussed by DiSalvo in his work could be described a working broadly within art, science and technology, and as experimental. Here we see artists and creative works engaged in a form of critical dialogue, which is beyond that which is possible within commercial design or purely technically focused research.

5   Art Science and the Experimental?

What confers [art] with its perennial possibility of eclipse is its function of rupturing with forms and significations circulating trivially in the social field. [...] Art confers a function of sense and alterity to a subset of the perceived world. The consequence of this quasi-animistic speech effect of a work of art is that the subjectivity of the artist and the 'consumer' is reshaped. The work of art, for those who use it, is an activity of unframing, of rupturing sense, of baroque proliferation or extreme impoverishment, which leads to a recreation and a reinvention of the subject itself. (Guattari 1995, p. 130)

In addressing the subject of art and science it is always tempting to provide a handy definition, yet any such definitions are fraught as both art and science are heterogeneous fields of practice, with their own contestations and deliberations about what constitutes art or science. As Wilson notes, the last century of Art history "has left the philosophy of art in turmoil"—making it "difficult to achieve consensus on a definition of art, the nature of the aesthetic experience, the relative place of communication and expression, or criteria for evaluation." Over the same period science as a field has gone through a series of what Kuhn (1970) describes as paradigm shifts. The claims and assumptions of science have been critiqued and questions by numerous authors, including Polyani (1964); Hacking (1983); Feyerabend (1985); Winner (1986)—foregrounding in many instances the social and human dimensions. As a result both within science, and in studies of science, there remains contestation regarding key questions of methodology, epistemology and ontology, which makes it difficult to resolve any shared consensus. More recently authors like Pickering (2010); Haraway (1998) and Lenoir (1998), have argue that scientific research produces highly situated knowledge and can be understood as cultural construct, rather than straight theory or facts. Thus Wilson (2002) comes to the conclusion that both art and science "make questionable truth claims and attempt to create privileged positions, but in reality participate in the system of symbols and narratives that shape the culture" (2002, p. 19).

1. Our apprehension of the world is active, not passive, and art displays an emergent apprehension.
2. Art is only incidentally and not essentially aesthetic. Art is concerned with every kind of value and not particularly with beauty.
3. Art interrogates the status quo; it is essentially, and not incidentally, radical.
4. Art is experimental action: it models possible forms of life and makes them available to public criticism. (Brook 1974)

For Brook (2012) experimental art is a form of "memetic innovation," a view that draws our attention to the way experimental art operates within the social field. While Pickering (2010) describes science as an encounter where machines, instruments, facts, theory, human disciplines of practice and social relationships are intertwined." If we are to appreciate art as social relation (Carter 2004), experimental art can be seen as a similar encounter—one that Brooks describes as emergent, radical and interrogating the status quo. Such practices for Guattari (1995, p. 130) involve a "rupturing of forms and significations circulating in the social field ... lead[ing] to a recreation and a reinvention of the subject itself" (1995, pp. 130–131). Thus experimental art is not an experiment conducted to produce singular truth, or falsify a hypothesis, as if operating in a world of scientific realism. Instead experimental art questions the assumptions of both art and science—which includes the logics of experimentation—through the way it explores the possible. This occurs in public, becoming part of a complex ecology of relationships—taking the form of a radical experiment.

Both science and art are emergent in nature, Pickering (2010) described science as an emergent practice occurring in "real-time", while Brooks' describes experimental art as involving an "emergent apprehension". So it not surprising that experiment and experience share similar etymological origins—both derived from the Latin 'experiens', meaning to "to try out". The definitional difference here is that experiment is intransitive, while experience is direct. We experience the world first hand, yet experiment upon or on the world. The emergent, 'real-time' nature of science suggests that this distinction is not so clear, and foregrounds the experiential nature of experiment. In a similar way, Feyerabend suggests that when considered from a cultural point of view—we mistake the structures of Being with the way "Being reacts to human interference" (1996). From this perspective we can come to an understanding of experimental art as a form of experiment that is designed to explore the meaning and structures of Being, by creating artistic/ aesthetic social experiences, experiments in which the audience plays a direct role.

A further decomposing of *experiens* reveals the prefix ex, meaning "out of", and the suffix periri', which refers to trial, peril, thus involving risk. It is this notion of risk that draws us to another similarity and difference between art and science. An experiment involves risk, without risk is it an experiment? Both artists and scientists take risks when they make commitments to their creative works, and hypothesis/ theories respectably. Both commit to an uncertainty, and once published, put themselves on the line publically for their work. As will be seen in the examples discussed, the risk of the experiment of art-science, is ever present but is transformed through the movement out of the research lab and into the public.

Art and science are full of contradictory theories, it is however easier to reconcile ambiguities within the arts, than science. Artists trade in metaphors, analogies, poetics, signs and symbols, all of which have no singular or fixed meanings, and in many instances the significance of a work lies in its ambiguity, or the complexity of multiple readings. Within science the most significant contradictions emerge from the view that science constructs the reality it studies. This is either through the very act of observation, for example in high-energy physics where particles of interest are created, as opposed to being found, within monumental experimental apparatus. Yet, as authors like Pickering, Freyebend and Ascot show, we do not need to turn to quantum mechanics to see these contradictions and tensions within science. Such philosophical debates may prove concerns within science, yet Ascot suggests it is artists who are "particularly responsive to the idea that nature is constructed" (2006, p. 9)—due to the fact they primarily deal with metaphor and other forms of ambiguity and uncertainty.

In science the experiment plays a pivotal role in theory generation (or falsification). Questions about the epistemological and ontological implications of experiments, and experimental apparatus, have been motivation for paradigm shifts over the history of modern science. Yet there prevails a view that theory is more important and separate from the messy embodied material real world in which experiments take place, and the tacit and practical knowledge that is integral to the success of any experiment. As such experiment is caught within a dichotomy—between theory (*theoria*) and practical skill (*techne*). This Feyerabend describes as a "conflict between a real but hidden world and a sham world that is accessible to humans" a conflict that he argues can be "found in all areas of human endeavor."

(1996) However, readings of the history of science—from primarily observational origins of Aristotle to Galileo, to the classical macroscopic experiments of Bacon to the modern microscopic experiments—have shown that the assumed relationship between *theoria* and *techne* are highly contestable. Sennett (2008) presents an analogous argument in his work the *Craftsman*, when he sets out to liberate the "practical man or woman at work" from the stereo type of *Animal laborans*. This divide between technical skills and imagination or higher-level activities attributed to *Homo Faber*, is for Sennett an artificial one. Central to both arguments is a complex interplay between *techne* and *theoria*—which revolves around the practice of making "artefacts."

> My basic image of science is a performative one, in which the performances—the doings— of human and material agency come to the fore. Scientists are human agents in a field of material agency which they struggle to capture in machines. Further, human and material agency are reciprocally and emergently intertwined in this struggle. Their contours emerge in the temporality of practice and are definitional and sustain one another. Existing culture constitutes the surface of emergence for the intentional structure of scientific practice, and such practice consists in the reciprocal tuning of human and material agency, tuning that can itself reconfigure human intentions. The upshot of this process is the construction and interactive stabilization of new machines and the disciplined performances and relations that accompany them. (Pickering 2010, p. 21)

It is the assumed position of the instrument within this performative doing that is science, specifically the experiment, which forms a focal point for critiques of artistic methods within a scientific paradigm. The concern expressed by many artists is that within a scientific paradigm there is the risk of art becoming instrumentalised. To be instrumentalised, Lelas says, is to "be eliminated, or at least transparent, something that leaves no trace." (1993) Thus for artists the issue is that when art is viewed as a research instrument, an experimental apparatus, employed as a method of data collection, to either generate theory or falsify hypothesis, it becomes subordinate to science and loses its value as art, as memetic innovation. Worst of all it becomes transparent, and can be eliminated.[10] Carter argues that "to conceive of the work of art as a detached datum is to internalize a scientific paradigm of knowledge production", which is "wrong for science" and fails to understand art as a social relation (2004, p. 10). The suggestion that this is "wrong for science" takes on multiple meanings when read in light of the previous discussions regarding the philosophy of science. On one hand the suggestion is that the methods of art are not suitable instruments for scientific research, and on the other hand the view that any instrument provides transparent access to nature is highly contested.

As we have seen in even a brief outline of the contestations in and about science, the (experimental) instrument cannot be eliminated so easily. The instrument is both the lens through which we discover the universe, and the machines within which we

---

[10] A similar argument is made in relation to art within industry innovation pathways, where creativity and artistic practices become an input which is easily instrumentalised in the logics of innovation, and as a result rendered invisible, or "eliminated".

create the universe. It has been brought into the world through the human hand. This temporal, emergent and performative process of making, and experiencing, leads to new ideas, new theories, that in turn cycle back influencing what we make. The things we make are both the result of theory, and deeply involved in the production of theory. This is a form of ontological designing; a material thinking; a performative materializing practice (Bolt 2001, 2004), a performative "doing" involving human and material agencies akin to Pickering's (2010) description of science. This observation reveals the importance of both the artifact, and the practices of making, within processes of knowledge creation, which is a required movement if we are to develop an appreciation of experimental art-science projects; and artistic research, as a recognized research methodology that can be described in respect to science.

There is one aspect of experiment, which marks normal science and art as different, which has not been addressed directly and is important to appreciate the methodological implications of experimental art-science. In science, experiments do not take place in "public"[11] and are not an end in themselves. Experiments are part of a larger process aimed to generate new knowledge that is presented to the world in other forms—research publications, theoretical constructs, patents and so forth. In contrast, the experiment of experimental art directly involves a social relationship rendered in public. The artwork, a performative experiment, is in itself an expression of knowledge. It is not simply about aesthetics or beauty, but is valued by the way it generates new meanings—memetic innovation. In this respect the experiments of art and science vary in a significant way.

This difference is in some countries recognized in research policy. In the Australian context, creative outputs can be recorded as research outputs in themselves, and universities across the country count creative outputs as part of their research collections. If this were to be the case for science—the artifacts created by and through science could in themselves be counted as research outputs—an experimental instrument or disease resistant crop for example could be presented as expressions of new knowledge. However, this would undermine the methodological significance of the scientific paradigm. That is, the ability to repeat, prove or disprove theory, and the associated processes of dissemination and critique—which in science takes the form of peer review and publication. In other words, it is in this way we come to a central difference between the two domains—that is the way knowledge is expressed, and value ascribed to works of art and science.

Art-science projects however find themselves in an unusual place within this landscape—and it is the projects that take on the form of public experiment—or experimental art—that "reconfigure the objects both of art and of scientific research" (Barry et al. 2008). Such works are not about only communicating scientific theory to a public, or visualizing data, but are instead closer to forms of scientific experiment. However, they take place in public, and often directly involve the public. Born and Barry use the example of de Costa's *Pigeonblog*[12] – a work which involves

---

[11] This is with the exception of some forms of citizen science, and the current movement towards "removing the walls" of the science lab that can be seen in many institutions.

[12] See the project web site: http://beatrizdacosta.net/Pigeonblog/statement.php.

gathering data about air pollution, and as an example is hard to easily describe as either art or science. The work makes pointed political and social commentary, questions the nature of scientific measurement, and is also cited in the early literature on the internet of things.[13] De Costa's work, like many art-science projects, is neither art nor science as commonly understood, it is instead "a social public experiment between humans and non-humans" (de Costa 2006). Such experiments Barry et al. see as "forge[ing] relations between new knowledge, things, locations and [people] that did not exist before" (2008).

6 Australian Context

In the late 90's Shaw suggested that there is limited evidence of "artistic work directly influencing science" (Shaw 1998, p. 165), and this concern remains current today as it was in the 1990s[14], with artists commonly finding it difficult to creating meaningful art science collaborations (Ox and Lowenberg 2013). Despite this, there is a small but surprising range of examples from which to draw inspiration. It is the drive of practitioners, described as "creative interdisciplinarians" (Wilson 2002), that has lead to the growth and development of Art Science as a field of practice. As a result there has developed an increasingly sophisticated and diverse species of art-science project, and an associated body of theory/ knowledge about such projects. In this final section of the chapter we will look at two examples from Australia, which are both recognized nationally and internationally as leaders in development of art-science projects and collaborations, and advancing transdisciplinary work. These two examples are *SymbioticA*, a bio-art research center at the University of Western Australia, and the Synapse program run by the Australian Network for Artists and Technology (ANAT), based in Adelaide.

6.1 SymbioticA—Experimental Bio-Art Practices

Many of the collaborations through the 1990s focused on "new media" and computational forms. Today art-science projects span the full spectrum of scientific research. One of the more challenging art-science fields of practices is that known as bio-art, which brings together artists and the biological sciences, as seen in the transgenic and living works Eduardo Kac. In Australia the leading site for this bio-art of work is *SymbioticA*[15]. While there are other pockets of work being conducted

---

[13] For example Julian Bleeker's "A Manifesto for Networked Objects—Cohabiting with Pigeons, Arphids and Aibos in the Internet of Things" (2006) See http://dm.ncl.ac.uk/courseblog/files/2010/04/whythingsmatter.pdf.

[14] See for example Ox (2013) paper "What Is the Challenge of Art/Science Today and How Do We Address It?".

[15] See the SymbioticA web site: http://www.symbiotica.uwa.edu.au/.

across the country[16], *SymbioticA* is by far the most concentrated and internationally recognized center, with a stated focus on "enabling artists and researchers to engage in wet biological practices" (UWA n.d.) within a science department. The center has since its inception supported numerous artists and projects through a program of residencies, education and research.

In this setting the tools and technologies of science are not simply used as a resource within a creative practice, nor do they form some distance object of commentary. The proposition that the tools and technologies of biological sciences could be raw materials, or resources, within a creative process is deeply problematic. This problem is not limited to artists working in biological materials in an experimental setting, but is the ethical territory all researchers in biotechnology operate within, and *SymbioticA* does not shy away from directly engaging in the related debates. They advocate that it is only through "experiential practice" within a scientific setting, that it is possible to develop an "understanding and articulation of cultural ideas around scientific knowledge." (UWA, n.d.) This is seen as being important for "informed critique of the ethical and cultural issues of life manipulation." On the surface this suggests the work follows a logic of accountability, where artists are engaged in work that in some form is aimed at holding science to account, and representing this to a boarder public. However, upon close inspection there is also an ontological logic, which is *"less with making art or science accountable or innovative than with altering existing ways of thinking about the nature of art and science"* (Barry et al. 2008) *This* can be seen in both sustained commitment to biological arts within a wet lab, and the programs of education, research, public presentation and scholarly publication. Similarly this program of activity suggests a mode of operation that cannot be easily reduced to antecedent disciplines. Instead it appears focused on developing the emergent field of bio-art, through a contestation of the assumptions implicit in the respective domains of art and science.

To provide an example we will draw a case study from one of the projects from *SymbioticA*—The Tissue Culture and Art Project[17] (TC&A) lead by Oron Catts and Ionat Zurr, which is described as "exploring the use of tissue technologies as a medium for artistic expression." (Catts and Zurr, n.d.) The project is however not simply focused on formal exploration of a particular artistic material, but instead is interested in "new discourses and new ethics/epistemologies" of the semi-living and "the contestable future scenarios they present". This is seen in the range of artworks that are part of the TC&A, one of the more poignant of which is *Victimless Leather*[18], which will be discussed below. *Victimless leather* is an excellent example of the ways in which Art-science operates – and provides an illustration of how concepts of instrumentation and experimentation play out within a creative work, and in an exhibition setting. *Victimless leather* is a small stitch-less jacket grown from immortalized cell, cultured and grown over a biodegradable polymer matrix,

---

[16] For example Dr Svenya Kratz who completed her practice-led PhD working at the Institute for Health and Biotechnology Innovation at the Queensland University of Technology.

[17] See the *Tissue Culture Art Project* web site: http://tcaproject.org/.

[18] See the *Victimless Leather* web site: http://www.tca.uwa.edu.au/vl/vl.html.

Fig. 2.1 *Victimless leather.*
(Oron Catts and Ionat Zur
2004)

described as problematizing the garment by making it semi-living (Zurr and Catts 2003). When exhibited, the semi-living jacket is presented along with a bioreactor, required to keep it "alive", and with each exhibition there is the associated performances of care—"feeding" during the exhibition, and ultimately killing of the work at the end (Zurr and Catts 2003). In this way the work not only employs the aesthetics of science, but is a functional example of the technology and practices of a tissue culture laboratory, and directly embodies the associated ethical dilemmas (Fig 2.1).

Victimless Leather appears to have come straight out of the "wet lab" into the gallery; with the small jacket grown from living cells only part of a whole. Its presence, and "semi-living" status, is only possible by way of the experimental setup designed to keep it alive. In this respect the work takes on the form of a public experiment, not an experiment the public partake directly in, but an experiment that unfolds in the public as opposed to behind closed doors. Like any experiment it involves risk, the risk of the experiment failing, the semi-living work "dying" prematurely through contamination, or through a lack of "care" in the practices required to sustain its living status. This risk, and the precarious and fragile existence of the living jacket, draws our attention to questions of care and responsibility. The care required to sustain an artwork made from living cells, and responsibility for its continued life. At the same time the work critically questions our existing relationship to clothing made from the skin of animals by suggesting a possible "victimless" future. But do such issues require an art-science response, or rephrased, why is an art-science project the most potent approach for addressing these issues? What is most important in addressing this question is that the artists are working with specific material technologies, which would otherwise be rendered as experimental apparatus—for data collection and writing about—but not ever published or circulated in themselves.

While the aesthetics of the biology experience, transparent glass, fluids, messy growths of cells, are present, it is only by way of an appreciation of the science and the associated ethical debates that the significance of the work becomes apparent. It is far more than an artifact for the polite, yet squeamish viewing, of gallery audiences. It is both a bringing into the world a creation, and a bringing forth a semi-living form created from cells—it is far from "natural"—yet we are squeamish because it is made from living cells, and presented with the apparatus that life support. Thus the material technologies of the biology wet lab are presented as art object within a gallery setting—constructed as an experience in order to generate thought and discussion. It is both an art object, an exploration of the formal properties of a "semi-living" material, as well as a real time experiment using contemporary scientific techniques. It thus questions the status quo and unsettle the normative assumptions of science and art and in the process become part of an emerging new field of bio-art.

Here we see a cue to the transdisciplinary nature of the project—in that what is presented as an artwork, is the result of the combined artistic and scientific disciplines thus transforming the "relations between artists and scientists and their objects and publics." (Barry et al. 2008) Through the practice of making the work, its exhibition, and writing about the works, the artists engage in a critical dialogue about the intellectual, ethical and political limits of biotechnology as a science, as well as existing normalized practices within society. This echoes the approaches of critical design[19], which challenge the normative practices of design and instead employ speculative design fictions to engage in a critical dialogue about topical issues. Yet were critical designers create non-functional fictional designs in order to critique issues related to science or new technologies—the work of TC&A goes one step further in that it employs the very science it is critiquing. It is both experimental art, and science experiment—yet cannot be comfortably understood as either. It is this inability for the work to be evaluated, or understood, from the perspective of existing disciplines that makes it of interest. To apprehend its value requires a reading of not only the respective disciplines of art and science, but the emerging field of bio-art.

Victimless leather is just one of over a dozen projects, each of which explore tissue cultures as artistic material, and at the same time engage in a similar critical dialogue. For example *Semi-living Steak* (2000) and *Disembodied Cuisine* (2003)

---

[19] Critical Design is best captured in the practice of Dunne and Raby (1999; 2001).

both of which explore the growing of meat for human consumption, research which is today, in 2013, only just reaching pre-commercial stages (Post 2013). Beyond the work of Catts and Zurr, *SymbioticA* runs an artist residency program, which since 2000 has supported over 60 artists from around the world.[20] As such *Symbi-oticA* is an important international example of transdisiplinary art-science, which is advanced through a combined program of education[21], workshops, symposium, exhibitions, research and artist residencies. When read in combination this provides one possible model for fostering, encouraging and empowering a new generation of "well-rounded" artists, scholars and students, through unconventional and creative application of science.

6.2 Synapse

> Collaboration between the arts and sciences has the potential to create new knowledge, ideas and processes beneficial to both fields. Artists and scientists approach creativity, exploration and research in different ways and from different perspectives; when working together they open up new ways of seeing, experiencing and interpreting the world around us. (ANAT n.d.)

The second example is the ANAT synapse program, which is aims to support collaborations between scientists and artists through the combination of a residency program, an online database of art-science practice (http://synapse.org.au) and an Australian Research Council (ARC) Linkage program that provides support for longer term projects that are developed through the initial residency.[22] Over the last 10 years the residency program has placed artists in close to 20 different research centers across the country, and internationally, in a disciplines ranging from Astrophysics to Synthetic Biology. A few examples of residencies supported through Synapse highlight the diversity of art-science partnerships: Chris Henschke at the Australian Synchrotron in Melbourne, Erica Seccombe in the Department for Applied Mathematics at the Australian National University, Robin Fox at the Bionic Ear Institute and George Pookhin Khut at The Children's Hospital Westmead. While each artist supported employs different tactics for engaging with science, there is a requirement for a joint application where there is a commitment to the collaboration from both artist and scientists. The program also has an explicit research focus, and is designed to allow for artists to immerse themselves within the science setting, and for the partners to develop an understanding of each other's respective practices. From this perspective Synapse is framed to foster transdisciplinary practices within

---

[20] For a list of Artists see http://www.symbiotica.uwa.edu.au/residents.

[21] *SymbioticA* run undergraduate and postgraduate courses on bio-art and Art Science practices, which are detailed at the following URL: http://www.symbiotica.uwa.edu.au/courses.

[22] This run through the Australian Research Council (ARC) Linkage grant scheme, which involve research conducted with an industry partner. For the Synapse Linkage scheme the industry partner is the Australia Council for the Arts. http://www.australiacouncil.gov.au/artforms/experimental-arts/opinion_piece._synapse_sharing_partnerships.

the context of science facilities—and in many instances artists engage in scientific research work as part of their residency, a strategy aimed at supporting longer term projects, which extend beyond the initial residency. Not all supported Synapse projects have gone beyond the initial residency; however there are several very notable ongoing projects that have emerged from Synapse residencies. Below we will mention two of note.

The first example, which will be discussed very briefly, is George Pookhin Khut's work with Dr Angie Morrow on the *BrightHearts* project[23]. The project is a synthesis of Angie's clinical knowledge and experience in medical research, and George's focus on the body combined with a background in research and "human-centered design methods and values", which has led to a relatively novel approach to the problem of "managing pain and procedure-related anxiety experienced by children" (Khut et al. 2011) Beyond being simply a medical research project, the resulting iPad application developed through *BrightHearts* research has been exhibited within gallery settings in exhibitions which span Art[24] and Design[25]. The work has produced several research publications, and has been recognized with an Australian Business Art Foundation (ABAF) award[26], and the Queensland New Media Art Award[27]. This range of outcomes demonstrates the value of the project across art, design and medical research and shows how such projects can lead to multiple outcomes without being subsumed by, or in service of the logics of either partner discipline (Fig 2.2).

The second example, which will be discussed in more detail, is the collaboration between Mari Velonaki, and roboticists David Rye, Steve Scheding and Stefan Williams at the Australian Centre for Field Robotics at the University of Sydney[28]. This collaboration developed from an initial residency, which to a 3-year Synapse Australian Research Council (ARC) Linkage grant to develop the *Fish Bird* project. The work *Fish-Bird: Circle B—Movement C*[29], which is one of a series of works, involves two autonomous robots in the form of wheel chairs, Fish and Bird, who are in love but cannot communicate directly. Instead they communicate with each other and an audience via movement and text (Velonaki 2008b).

Presented publically—in gallery settings—*Fish-Bird: Circle B—Movement C* is a unique artistic work, and experiment in robotics, one which has allowed Velonaki and collaborators to explore the central problems of the Center for Social Robotics,

---

[23] See the *Brighthearts* web site: http://georgekhut.com/brighthearts/.

[24] For example *Brighthearts* was included in, and won, the 2012 New Media Art Awards at Gallery of Modern Art in Brisbane Queensland. See http://www.qagoma.qld.gov.au/exhibitions/past/2012/national_new_media_art_award_2012.

[25] The work has also been included in *CUSP; Designing the Next Decade* curated by the Australian Center for Design. Similarly the work of Mari Vilonaki discussed in this paper is also included in the same exhibition. See http://www.cusp-design.com/.

[26] See the Arts and Health Foundation Award 2012 http://www.creativepartnershipsaustralia.org.au/arts/awards/2012-abaf-award-winners.html.

[27] Ibid. 25.

[28] See the Australian Center for Field Robotics web site: http://www.acfr.usyd.edu.au/.

[29] See the artists web site: http://mvstudio.org/work/fish-bird-cicle-b-movement-b/.

Fig. 2.2 *Brighthearts*. (By George Poookhin Khut. Photograph by Julia CharlesVelonaki 2012, 2004, 2011)

which formed in 2006 within the Australian Centre for Field Robotics. Specifically the aim was to "study human-robot interaction in social environments"—which they describe as requiring "a multidisciplinary understanding of the science, technological sociological and cultural dimensions of human robot interactions" (CSR, n.d.). In this way *Fish-Bird: Circle B—Movement C* becomes an experiment played out in public, having been show around the world and encountered by tens of thousands of people, a field trial of a scale that would be difficult without the artistic setting. However, the experimental approach goes a step further than *Victimless Leather*, in that it directly involves the audience within the dynamics of the experiment, through their interaction with the work in a direct manner. The audience becomes part of an experiment, both as participants in a study of human-robot interactions, and as actors within an unfolding robotic love story (Fig 2.3).

Since forming as a group in 2006 Velonaki and her collaborators have won research and arts funding to support a number of similar projects, which have been exhibited internationally, and also resulted in publications across art and science. Of note are developments that come out of this body of research including two PhD theses. One studying artificial skin and the interpretation of touch in human-robot

Fig. 2.3 *Fish-Bird: Circle B*—Movement C. By Mari Velonaki 2004

interaction (Silvera Tawil 2012); and a second focusing on the psychophysiological correlates of emotive and cognitive variables in computer-based tasks (Brown
2012). Similarly, through the work of the center, Velonaki and collaborators have
refined a methodological approach to support their research. This focuses on multi-
objective evaluation (Velonaki 2012), which is framed from a multi disciplinary
perspective, as opposed to inter or trans. This appears a strategic choice; to design
projects to produce outputs that have meaning within respective disciplinary contexts, yet work together as a whole. This approach may in part be driven by an environment that is structured along the lay lines of disciplines, both the university and
funding bodies. For example both arts funding and research funding is commonly
structured around disciplines—with reviewers for applications selected based on
disciplinary expertise. The one exception to this is the Experimental Art program
run by the Australia Council for the Arts.

In the paper *Shared Spaces: Media Art, Computing, and Robotics*, Velonaki,
Scheding, Rye and Durrant-Whyte, note that they came together around a shared
interest in "the creation of human-machine interfaces" Velonaki 2008a, b. This is of
specific note in the context of interdisciplinary fields, as the field of Human-Computer Interaction (HCI) has undergone several transformations as a result of interdisciplinary encounters; from cybernetics and man-machine interaction, to HCI and
Interaction design, to the contemporary formulation of experience design, which
has a curious echo of experimental. So while their methodological approach articulated as multi, we see in this example the emergence of an approach which involves

a "synthesis of research at the stages of conceptualization, design, analysis, and interpretation by integrated team approaches" (Hadorn 2008. Cited in Ertas 2010)

In Falk's 2011 case study Rye described the significance of the work of the CSR as the way it "declares the importance of an area, which is really the non-technological, non-scientific, non-engineering influences on robotics". In this way the CSR and its approach prefigure the movement of robotics and robots, out of the lab and into the world. This can be seen as analogous to the transformations of HCI and Computer Science as computers became ubiquitous through the 1990s to the 2000s. As a result there is a demand robotics (and robots) become a focus for study and research, where "psychologists, anthropologists and cultural theorists who can interpret relational dynamics and data" (Falk 2011) come with scientists and engineers to address the challenges and problems resulting from the increasing use of a new technology within society.

In this example art gallery becomes an experimental space where the human audience plays a role in a multi objective evaluation framework. This renders any close reading of the particulars of the project difficult, as one needs to be appraised of the significance the knowledge outcomes across different disciplines. What is of specific interest in this example and in the context of this chapter is the way Mari's creative practice and robotic works, *Fish-Bird* and more recently works like *Diamandini*, provide a point of focus for a sustained research agenda, and a unique opportunity to study such human interactions with robots outside the lab. As an example of art-science we see the artwork become instrument, experiment, creative expression and experience, without becoming invisible through this process. It is the complex interrelations that form around the creative work, which provide the potential for new knowledge far beyond technical innovation alone (Fig. 2.4).

Falk presents the value in terms of the logic of innovation: "This story captures a recurring theme in creative innovation: collaborations foster interpersonal relationships that can kick-start long-term innovation paths and engagement across industry lines." (Falk 2011) This highlights the importance of such collaborations and mechanisms for fostering relationships and sustaining long-term projects. However these projects cannot be viewed through the logics of innovation alone—for it is the "logic of ontology" which is key to transformation, as is seen in the evolution of HCI as a field of research and the resulting ways this has transformed our relationships with, and understandings of technology. Similarly, there is a two way movement between art and science which is seen in Velonaki's recent move to the University of New South Wales to lead the Creative Robotics Laboratory[30] with the National Institute for Experimental Arts.[31] Here research begun within a more traditional robotics research center, conducted within a multi-disciplinary frame, has evolve into a program within a research center that like *SymbioticA* has been established with a focus on transdisciplinary experimental arts practices.

---

[30] See the Creative Robotics Lab web site: http://www.niea.unsw.edu.au/about/niea-groups/creative-robotics-lab.

[31] See the National Institute for Experimental Arts web site: http://www.niea.unsw.edu.au/.

Fig. 2.4 *Diamontini*. (By
Mari Velonaki 2011)

From even this short investigation of just two of the art-science partnerships supported through Synapse there emerges a picture of the potential for such transdisciplinary experimental practices. However, Synapse as a program stands in the face of a system of funding and reporting which remains polarized along disciplinary lay lines, and often driven by the demands for immediate outcomes. This second point on immediate outcomes, with projects framed in periods of typically 12 months to 3 years, undervalues the importance of time in the development of transdisciplinary practices. Successful Synapse projects, like those discussed above, are only possible through extended work, where there is time for the partners to develop a "theoretical, conceptual, and methodological reorientation" (McMichael 2000) seen in mature transdisciplinary practices. Upon close inspection Synapse may prove to be an international benchmark in this respect.

7 Conclusions

To conclude we return to da Vinci, and a view of his work as a polymath which questions the contribution he made to scientific knowledge. Of Leonardo Carrier says "[his] art is great, but his studies of science and technology are of interest only to intellectual historians." (2008) For Carrier it is not da Vinci's contributions to science that are significant, but instead his "fascination for the relationship of art and science" (2008), as expressed not just though his artworks, but through his journals and sketches. It is this interest in the relationship between art and science that we suggest is more important than dreams of a contemporary "Renaissance man". What this suggests is a trans disciplinary approach—one that Bennett describes as engaging with, and motivated by, "external conditions or problems [and a] conviction that disciplines [should] not have proprietary rights over their domains" (Bennett 2012). For it is from the places where disciplines come together and interact, that new meaning emerges.

As we have seen such forms of practice do exist, however the question remains as to how these practices—often grounded in collaborations—are formed, fostered and supported. And more specifically how such an appreciation for Art **and** Science is realized in curriculum. With the exception of a few exemplars, the mainstream body of university education remains polarized around disciplinary "silos". The structures and logics of institutional organization, funding and reporting, form an inertia pull towards disciplines that needs to be questioned, and resisted—not just by faculty; but by students themselves. In the face of looming change in the university sector internationally—driven by developments in information and communications technology that are transforming the way education is produced and delivered—there is no more potent moment for the radical rethinking of disciplinary based programs.

In response to the question posed by *Digital da Vinci* we suggest a subtle reformulation—how can we encourage the same type of interest, or fascination in our scholars and students for both art and science as we see in Leonardo da Vinci? From surveying the state of art/science practice over the last 50 years what is clear is that this is not necessarily a problem of the arts. What is more urgently required is for the science, technology, engineering and mathematics disciplines to genuinely see the substantial value art and creative practices bring to their respective fields. Here a new formulation has emerged which includes art which is gaining traction in internationally.[32] It is this recognition of the value of art, beyond a subordinate mode of operation framed by the logics of pre-existing disciplines, which is central to addressing the challenges posed by this text—*Digital da Vinci.*

---

[32] See for example Roger Malin's work Chair of Arts and Technology at the University of Texas http://www.utdallas.edu/atec/malina/ and the Science Engineering Art and Design (SEAD) developments http://www.utdallas.edu/atec/cdash/ and http://seadnetwork.wordpress.com/, the Leonardo Education and Art forum http://www.leonardo.info/isast/LEAF.html, the liberal arts and engineering programs at California Polytechnic State University http://laes.calpoly.edu/, and the Rhode Island School of Design's STEM to STEAM program http://stemtosteam.org/.

References

Ascott, Roy. 2006. *Engineering Nature: Art & Consciousness in the Post-biological Era*. Bristol UK: Intellect Books.

Australian Network for Art and Technology (ANAT). 2008. Synapse – Art Science Collaborations. http://www.synapse.net.au/. Accessed September 10, 2013.

Austin, J. L. 1962. *How to Do Things With Words*. London: Oxford University Press.

Bakhshi, Hasan, and Throsby, David. 2010. *Culture of Innovation*, London: National Endowment for Science, Technology and Arts (NESTA).

Bakhshi, Hasan, Freeman, Alan and Potts, Jason. 2011. *State of Uncertainty*, London: National Endowment for Science, Technology and Arts (NESTA).

Barry, Andrew. Born, Georgina, and Weszkalnys, Gisa. 2008. Logics of Interdisciplinarity. *Economy and Society* Vol. 37, No. 1: 20–49. doi:10.1080/03085140701760841.

Barthes, Roland. 1973. *Mythologies*. St Albans, Herts.: Paladin.

Beddard, Honor. 2009. Computer Art at the V&A. *V&A Online Journal* no. 2. http://www.vam.ac.uk/content/journals/research-journal/issue-02/computer-art-at-the-v-and-a/. Accessed September 10 2013.

Bennett, Jill. 2012. What is Experimental Art. *Studies in Material Thinking*, Vol. 8. http://www.materialthinking.org/papers/88. Accessed 8 September 2013.

Bleecker, Julian. 2006. A Manifesto for Networked Objects—Cohabiting with Pigeons, Arphids and Aibos in the Internet of Things. http://dm.ncl.ac.uk/courseblog/files/2010/04/whythingsmatter.pdf. Accessed 8 September 2013.

Bolt, Barbara. 2001. Materialising Practices: The Work of Art as Productive Materiality. PhD thesis, Murdoch University.

Bolt, Barbara. 2004. *Art Beyond Representation: The Performative Power of The Image*. London: I.B Tauris & Co Ltd.

Born, Georgina, and Barry, Andrew. 2010. Art-Science. *Journal of Cultural Economy* Vol. 3 No. 1: 103–119. doi:10.1080/17530350003617610.

Bourriaud, Nicolas. 2002. *Relational Aesthetics*. France: Les presses du réel.

Bridle, James. 2011. Waving at the Machines. http://www.webdirections.org/resources/james-bridle-waving-at-the-machines. Accessed 6 September 2013.

Bridle, James. 2012. "#sxaesthetic report". http://booktwo.org/notebook/sxaesthetic/. Accessed 6 September 2013.

Brook, Donald. 1974. About the Australian Experiemental Art Foundation. http://aeaf.org.au/about/aboutaeaf.html. Accessed 7 September 2013

Brook, Donald. 2012. Experimental Art. *Studies in Material Thinking*, Vol. 8. https://www.materialthinking.org/papers/101 Accessed 8 September 2013.

Brown, Paul. 2008. From Systems Art to Artificial Life Early Generative Art at the Slade School of Fine Art. In *White Heat Cold Logic: British Computer Art 1960–1980*, ed. Gere, C., Brown, P., Lambert, N. and Mason, C. Cambridge, Massachusetts: MIT Press.

Brown, I. 2012. *Cognitive context: exploring psychophysiological correlates of emotive and cognitive variables in computer based tasks*. PhD thesis, Australian Centre for Field Robotics, The University of Sydney.

Carrier, David. 2008. Introduction: 'Leonardo' and Leonardo Da Vinci. *Leonardo* Vol. 41, No. 1: 36–38. doi:10.2307/20206514.

Carter, Paul. 2004. *Material thinking: the theory and practice of creative research*. Melbourne, Australia: Melbourne University Publishing.

Catts, Oran, and Ionat Zurr. n.d. The Tissue Culture and Art Project. http://tcaproject.org/. Accessed September 10, 2013.

Catts, Oran, and Ionat Zurr. 2000. Semi-Living Steak. http://tcaproject.org/projects/victimless/steak. Accessed September 10, 2013.

Catts, Oran, and Ionat Zurr. 2003. Disembodied Cusine. http://tcaproject.org/projects/victimless/cuisine Accessed September 10, 2013.

Catts, Oran, and Ionat Zurr. 2004. Victimless Leather. http://tcaproject.org/projects/victimless/leather. Accessed September 10, 2013.

Centre for Social Robotics (CSR). n.d. About Us. http://www.csr.acfr.usyd.edu.au/about.htm. Accessed September 11, 2013.

Coessens, Kathleen, Darla Crispin, and Anne Douglas. 2009. *The Artistic Turn: a Manifesto*. Brussel: Leuven University Press, 2009.

Csikszentmihalyi, M. 1999. Implications of a Systems Perspective for the Study of Creativity. In *Handbook of Creativity*, ed. R. J. Sternberg. Cambridge: Cambridge University Press.

Da Costa, Beatriz. 2006. PigeonBlog. http://beatrizdacosta.net/Pigeonblog/index.php. Accessed September 10, 2013.

DiSalvo, Carl. 2012. *Adversarial Design*. Cambridge, Massachusetts: MIT Press.

Dunne, Anthony. 1999. *Hertzian Tales. Electronic Products, Aesthetic Experience, and Critical Design*. Cambridge, Massachusetts: MIT Press.

Dunne, Anthony, and Fiona Raby. 2001. *Design Noir: the Secret Life of Electronic Objects*. London: August Media Birkhäuser.

Ertas, Atila. 2010. Understanding of Transdiscipline and Transdisciplinary Process. *Transdisciplinary Journal of Engineering & Science* Vol. 1, No. 1:54–73.

Falk, Michael. 2011. Assembling Social Futures: The Centre for Social Robotics. *Creative Industries Innovation Centre*. http://www.creativeinnovation.net.au/Media/docs/CSR-4f2e1256-8ae5-4491-9680-5e04f3d69fe0-0.pdf. Accessed September 10, 2013.

Freyer, Conny, and Troika (Firm). 2008. *Digital by Design: Crafting Technology for Products and Environments*. London: Thames & Hudson.

Feyerabend, Paul. 1985. *Problems of Empiricism: Volume 2: Philosophical Papers*. Cambridge: Cambridge University Press.

Feyerabend, Paul. 1985. *Realism, Rationalism and Scientific Method: Volume 1: Philosophical Papers*. Cambridge: Cambridge University Press.

Feyerabend, Paul. 1996. "Theoreticians, Artists and Artisans." *Leonardo* Vol. 29, No. 1: 23–28. doi:10.2307/1576272.

Fry, Tony. 2009. *Design Futuring: Sustainability, Ethics and New Practice*. Sydney: University of New South Wales Press.

Gere, Charlie. 2008. New Media Art and the Gallery in the Digital Age. In *New Media in the White Cube and Beyond Curatorial Models for Digital Art*, ed. by Christiane Paul. Berkeley: University of California Press.

Guattari, Felix. 1995. *Chaosmosis: an Ethico-aesthetic Paradigm*. Bloomington, Indiana: Indiana University Press.

Hacking, Ian. 1983. *Representing and Intervening: Introductory Topics in the Philosophy of Natural Science*. Cambridge: Cambridge University Press.

Hadorn, H. G., Susette Biber-Klemm, Walter Grossenbacher-Mansuy, Gertrude Hirsch Hadorn, Dominique Joye, Christian Pohl, Urs Wiesmann and Elisabeth Zemp. 2008. Handbook of Transdisciplinary Research, Enhancing Transdisciplinary Research: A Synthesis in Fifteen Propositions. Chapter 29, p. 435.

Haraway, D. 1988. Situated knowledges: The science question in feminism and the privilege of partial perspective. *Feminist Studies* Vol. 14, No. 3:575–599.

Khut, George Poonkhin, Angie Morrow, and Melissa Yogui Watanabe. 2011. The BrightHearts Project: a New Approach to the Management of Procedure-Related Paediatric Anxiety. OZCHI Workshops Program, Canberra.

Kuhn, Thomas S. 1970. *The Structure of Scientific Revolutions*. 2nd ed. Chicago: University of Chicago Press.

Lelas, Srdjan. 1993. Science as Technology. *The British Journal for the Philosophy of Science* Vol. 44, No. 3: 423–442. doi:10.2307/688014.

Lenoir, Timothy. Ed. 1998. *Inscribing Science: Scientific Texts and the Materiality of Communication*. California, US: Stanford University Press.

Manovich, Lev. 2001. *Post-media Aesthetics*. DisLocations. Karlsruhe: ZKM / Zentrum für Kunst und Medientechnologie.

Manovich, Lev. 2003. Introduction. in *The New Media Reader*, ed. Wardrip-Fruin, Noah, and Montfort, Nick. Cambridge, 13–25. Cambridge, Massachusetts: MIT Press.

McMichael A.J. 2000. What makes transdisciplinarity succeed or fail? First Report. In *Transdisciplinarity: recreating integrated knowledge* ed. Somerville MA, Rapport DJ. Oxford, UK: EOLSS Publishers Ltd.

Mouffe, Chantelle, *Agonistics: Thinking The World Politically*. London – New York: Verso, 2013.

Nowotny, H., Scott, P., & Gibbons, M. 2001. *Re-thinking science: Knowledge and the public in an age of uncertainty*. Cambridge: Polity.

Ox, Jack, and Richard Lowenberg. 2013. What Is the Challenge of Art/Science Today and How Do We Address It? *Leonardo* Vol. 46, No. 1: 2–2.

Paul, Christiane. 2003. *Digital Art*. London: Thames & Hudson.

Paul, C. 2008. *New media in the white cube and beyond: curatorial models for digital art*, University of California Press, Berkeley

Pickering, Andrew. 2010. *The Mangle of Practice: Time, Agency, and Science*. Chicago: University of Chicago Press.

Popper, Frank. 1993. *Art of the Electronic Age*. London: Thames & Hudson.

Post, Mark. 2013 *Cultured Beef Project* Maastricht University. http://www.maastrichtuniversity.nl/web/show/id=6866536/langid=42 Accessed September 20, 2013.

Quaranta, Domenico. 2010. *Media, New Media, Postmedia*. Milan: Postmedia Books.

Rittel, Horst WJ, and Melvin M Webber. 1973. Dilemmas in a General Theory of Planning. *Policy Sciences* Vol. 4, No. 2: 155–169.

Rush, Michael, and Michael Rush. 2005. *New Media in Art*. 2nd ed. World of Art. London: Thames & Hudson.

Silvera Tawil, D. 2012. *Artificial skin and the interpretation of touch in human-robot interaction*. PhD thesis, Australian Centre for Field Robotics, The University of Sydney.

Sennett, Richard. 2008. *The Craftsman*. New Haven: Yale University Press.

Shanken, Edward. 2009. *Art and Electronic Media*. London: Phaidon Press.

Shaw, Jeffery. 1998. Convergence of Art, Science and Technology. In *Art @ Science*, ed. Editors Sommerer, Christa, and Laurent Mignonneau, 162–166. Springer.

Snow, C. P. 1964. *The Two Cultures*. London: Cambridge University Press.

The University of Western Australia (UWA). SymbioticA. http://www.symbiotica.uwa.edu.au/. Accessed September 10, 2013.

Waddington, C. H. 1968. New Visions of the World. *Leonardo* Vol. 1, No. 1: 69–75. doi:10.2307/1571907.

Waddington, C. H. 1969. *Behind Appearance: a Study of the Relations Between Painting and the Natural Sciences in This Century*. Edinburgh: Edinburgh University Press.

Wands, Bruce. 2006. *Art of the Digital Age*. London: Thames & Hudson.

Willis, Anne-Marie. 2006. Ontological Designing. *Design Philosophy Papers* no. 2.

Wilson, Stephen. 2002. *Information Arts: Intersections of Art, Science, and Technology*. Cambridge, Massachusetts: MIT Press.

Winner, Langdon. 1986. *The Whale and the Reactor: a Search for Limits in an Age of High Technology*. Chicago: University Chicago Press.

Winograd, Terry, and Carlos F. Flores. 1986. *Understanding Computers and Cognition: a New Foundation for Design*. Norwood, N.J.: Ablex Publishing.

Velonaki, Mari. 2012. Multi-Objective Evaluation of Cross-Disciplinary Experimental Research. *Studies in Material Thinking*, Vol. 8. https://www.materialthinking.org/papers/100. Accessed 8 September 2013.

Velonaki, M., Scheding, S., Rye, D. & Durrant-Whyte, H. 2008a. Shared spaces: media art, computing and robotics. *ACM Computers in Entertainment*. Vol. 6, No. 4: 51:1–51:12.

Velonaki, M., Rye, D., Scheding, S. & Williams, S. 2008b. *Fish-Bird: A perspective on cross- disciplinary collaboration IEEE MultiMedia*, January–March:10–12.

Chapter 3
Repeating Circles, Changing Stars: Learning from the Medieval Art of Visual Computation

Mine Özkar

1 Repetition: The Computable Goodness of Design

Good designs, very generally speaking, have a repetitive quality. Goodness in repetition has little to do with the viewer's comfort in receiving the expected. Rather, we appreciate repetition because it allows us to recognize—or even to think that we wondrously discover—the new and the different amidst similarities. Whereas repetition implies consistent relations of similar parts, differences challenge these relations and stimulate our interpretive capacity towards recognizing multiple, unique but still meaningful, wholes. Dialogues that arise from repetition and variation characterize a good design. The aim below is to draw attention to a centuries old visual design with a repetitive quality that resonates with computational iteration while finding its character in variations that result from seeking and seeing different relations.

Repetition is common to many art forms. The literary world offers, in widely available contemporary resources on grammar, composition and literary terminology, a broad range of technical categories of repetition as rhetorical device. These categories as well as notions of disordered rhythm and defamiliarization from early literary formalism (Shkolovsky [1925] 1991) are, to some extent, of interest to those who wish to articulate what repetition implies for the visual arts. Surprisingly, perhaps because of the wide range of its media, the artistic world does not present such a common analysis of repetition techniques. Nonetheless, repetition often finds its artistic counterpart in rhythm. And spatial rhythm, as an organizational phenomenon, is widely accepted as an experiential treat and invaluable trait in modernist art and architecture. Rasmussen's handbook to modernist values in architecture (1959) is a key resource for this understanding. There is also a considerable amount of current literature that recognizes rhythm in visual designs and architectural space. Writings range from rigorous scholarly approaches such as the comprehensive *rhythmanalysis* of urban environments (Lefebvre 2004), or discussions of the role of redundancy

M. Özkar (✉)
Istanbul Technical University, Istanbul, Turkey
e-mail: ozkar@itu.edu.tr

N. Lee (ed.), *Digital Da Vinci*, DOI 10.1007/978-1-4939-0965-0_3,
© Springer Science+Business Media New York 2014

in perceiving space (Lawson 2001) to popular architecture blogs illustrating desired architectural spaces old and new with compositional repetitions. Sometimes, even if outside the realm of music, definitions of rhythm include movement and time. Within the scope of this text, the only movement in time required to observe any rhythm is that of the viewer's gaze. Rhythm is referred to as any perceived totality of repetitions and variations in a finished visual design.

Talking about repetition is an essential part of contemporary architectural design education, especially at the foundational level. As aspiring designers learn about design thinking, they begin to understand how and why one establishes relations between parts towards creating the larger meaningful whole. When we ask why a part of design is so, answers expose its relations to other parts: the arches above the windows mimic the section of the ceiling vault. Sometimes answers refer to parts from a broader external context: the arch of the window is a structural requirement to span an opening that size.

In most good designs, we seek and find wholes. In designs where all relations are unique, a sense of whole is lost. Unique relations between parts are only appreciated when there are repeating relationships at other levels. For Pollock's paintings where individual paint drops seem to have been accidentally placed, we seek and appreciate a larger whole for the entire canvas, calling out visual patterns. Alternatively we refer to the entire set of the artist's similar paintings acknowledging his process of making. Pollock's bodily performances provide a consistent reference for the paint drops across all his paintings.

Repetition, as essential as it is for design, is not always straightforward to perceive and to conceive. Firstly, repetition of parts, which can be shapes as well as features or relations, is almost never monotonous in good designs. Parts do not necessarily repeat in exactitude to be deemed repetitious. Hence there is variation, and the notion of rhythm that encompasses both. The line between an exact copy and something similar is blurred. Our perception of what repeats helps us establish what parts are in the first place. Still, some parts may be deemed repetitious due to some of the smaller repeating parts within. Parts within wholes thus vary.

Let us assume that we generally perceive wholes, and may call them parts if we associate them with a larger entity. The repeating parts from different wholes help us relate these wholes to one another. Parts are similar at various scales and this similitude helps us connect parts to one another at these levels. Parts that are similar to one another seem to belong together. Secondly, a compositional whole does not begin and end in the object but expands to its environment also. For example, a soda bottle is a compositional whole in itself but also could be considered a part in a larger compositional whole when the consumer is holding it in their hand or it is among many similar bottles organized side by side on the market shelf. Contexts multiply our perception of part-whole relations.

Fig. 3.1 Various decompositions of a shaded circle

2 Variation: The Visually Computable Counterpart to Repetition

In design education at the foundational level, students struggle with making decisions. Creativity is difficult when one has not yet developed a sense of how to set up a temporarily constrained environment that allows you to make decisions, one at a time and with implications on another decisions. If one can control relations, it is possible to make choices that implicate the whole. Creativity in design is more dependent on context than its counterpart in the arts.

Conventionally we have come to understand computation as the act of counting identity relations, which also passes as copying or repeating a set relation. Stiny (2006) unconventionally reflects on this type of processes as zero dimensional because relations between parts are never questioned and set beforehand. Parts in digital computation are in fact primitives that cannot be divided into unprecedented smaller parts. Stiny alternatively draws attention to part relations that are technically called embedding. In higher-than-zero dimensional entities, parts are allowed, are infinitely many, and get decided on, temporarily, by the eye of the beholder. The eye embeds unprecedented parts in wholes. Decompositions change from person to person, from context to context, hence the potential for variation. Figure 3.1 shows alternative parts cut out from a circle. Once we identify parts as we see fit, we can count and copy them. In the third and fourth decompositions in Fig. 3.1, there are two of each unique part whereas in the fifth, there are six. There is yet another level of identity relations here. Although different in shape, these five alternatives are particularly derived from the same constrained decomposition structure that is the key player of the text below.

The repeated relations make design computable in the zero dimensional way. Nonetheless, good designs also display a variance among parts, and the creative idea often exists in recognizing this variance among similarities. The variance is not only to successfully address the multiple contexts inherent in any design problem. Limiting the discussion to abstract and visual designs only, it is also due to the very nature of visual thinking where the eye composes and decomposes parts and wholes in different ways, as seen in Fig. 3.1. The designer and the viewer appreciate the multitude of ways of relating to the work. Variation should be seen as opportunity to create. Following in Stiny's footsteps, the indeterminacy in the involvement of the eye indicates that design computability is visual. Being visually computable implies that seeing complements repetition.

As it turns out, the visual interplay of repetitive and varied parts in design is a centuries-old problem in architectural design. One instance, which will be the focus of this text, is the geometric patterns extending over the surfaces of Seljuk (and more generally medieval Islamic) architecture and are based on repeating circles. Signifying infinity and continuity, these designs display a strong repetitive quality. They are repetitive not only in the endless iteration of units, but also because they conform to a style by reiterating certain motifs. Similar forms across geography help us identify them as what they are but this repetitive quality is well balanced with pattern variety as most are unique.

The geometric ornaments in Islamic architecture show variation over time and geography. They cover a larger domain within which the star-shaped Seljuk patterns comprise a particular geographical and temporal frame. The patterns of concern in this paper are specific to the central and southeastern regions of Asia Minor and the culture of eleventh to thirteenth century Islam in that geography. Rather than being a historical study, this investigation focuses on visual composition in these patterns and its geometric construction.

3 Seljuk Patterns: Repetitions of Constraints and Variations Upon Sight

A uniform field of interlocking circles, drawn simply on any surface using a compass and nothing more, is the underlying guide for constructing various geometric relations of stars, convex polygons and straight lines. These shapes emerge when the artisan's eye connects intersection points with new lines. The circle grid, usually based on the repetitive relation of two circles that pass through the other's center, also known as *vesica piscis*, ensures the unity of the overall structure as well as the style that sets these ornaments apart from others. It also yields to creative variation as the artisan visually constructs the lines that eventually form unique patterns (Fig. 3.2). I argue that these patterns, with their constrained variety, illustrate the basics of good design, and are relevant for establishing a direct link between computation and good design.

Design problems and solutions are always unique as each design is ideally consistent and relevant in its context. In the particular case of the medieval Anatolian architectural ornaments, the motivation and the limitations of the design are very much defined by the social dynamics of the era (Mülayim 1982; Bakirer 1981). The times are the rise of philosophical Islam. Figurative decoration is not permissible, so ornamentation on stone portal facades or on wooden minbars from the interiors is only geometric. The common motifs are variations of what is known to be the "infinity motif" or the "star motif". In these motifs, large areas of stone or wood are covered with repetitive patterns extending to the edges.

The patterns are a part of the Sufi culture and constrained due to the expectations of the patrons and the belief system. They are also a part of a more elaborate scheme of abstract art that decorated mostly portals of public buildings, side panels

Fig. 3.2 Two patterns of common hexagonal and octagonal geometry and how they are constructed using circle grids based on *vesica piscis*. Steps are originally developed by Bakirer (1981) and redrawn here by the author

of tombs and grills used for privacy indoors. The architecture of the period, especially in the region, displayed a mystic symbolism in volumetric organization. The ornamentation supplemented that stereotomy. There is extensive literature on the Sufist meanings behind the ornaments (Ogel 1966) but these remain beyond the scope and interest of this paper.

Various inventories for geometric patterns have already been formed both from the art historian's and the mathematician's point of view (Critchlow 1976; Mülayim 1982; Bakirer 1981; Grünbaum and Shephard 1987, 1992; Lee 1995). Much has been said on the mathematics and computational aspects of tile-like patterns (Abas and Salman 1992; Kaplan 2000). More recently, Lu and Steinhardt (2007) have revealed the sophisticated geometry, particularly of aperiodic sequences in Islamic patterns in Uzbekistan, and renewed attention to these patterns in the Western world. Nevertheless, comparative to the studies on the mathematical and computational aspects of tiling, little has been said about what the eye is doing in these strongly visual designs.

Despite the common knowledge of the flow of mathematical and scientific knowledge from medieval Eastern manuscripts to the West, elaborate mathematics was not common knowledge among the medieval age craftsmen. They knew how to implement shape construction practically but not necessarily the mathematics behind it. Özdural (2000) documents the eleventh century carpenter's wondrous and visual ways of efficiently dividing and uniting wood panels without the use of any calculation device. Their skill was dependent on the learnt visual and spatial traditions of the craft. It is not clear whether it was mostly a draftsman who completed the designs on paper prior to the material application. Where the design is carried out makes a great difference to the nature of the process. Still, overlooking the technical and physical restrictions, this study focuses on the visual aspects. As discussed by Özkar and Lefford (2006), the style of these patterns is based on procedures and on how final shapes are constructed. Seeing sets the visual process behind these patterns apart from mere iterations and there is much to be said about the visual computation. Below, I make a case for the visual process that characterizes this style with regards to repetitions and variations.

The idea that these patterns are constructed on circle grids is evidenced in an earlier study on Seljuk ornamentation by art historian and architectural conservationist Ömür Bakirer (1981). Bakirer then reinstated Seljuk geometric patterns as formal and procedural investigations more than just mystic ornamentations or mathematical exercises. Her study is part of an inventory produced for a doctoral dissertation on bricklaying in Seljuk architecture with concentration on surfaces. Differently than most studies on geometric patterns in the Islamic world, she deals not only with abstract forms but also with how they are constructed. She is interested in understanding these constructions not being determined by rigid formulae but rather becoming in a process that adapts to the emerging geometry. Materiality and constructive knowledge of geometry is key to Bakırer's explanation. She extends her investigations with more case studies (1984) and provides evidence that artisans indeed carved interlocking circles on the stone on site (1999).

Although some aspects of Bakirer's approach are outdated, it still provides a story with evidence about how the craftsman applied the geometry on the stone. It also sheds light on how the designer may have drafted the design in a process of visual thinking. This approach is the basis for Özkar and Lefford's argument for process based understanding of style. Lu and Steinhardt's article (2007) similarly has a process based interpretation, but with a significant difference. They propose geometric tiling as the process behind some of the more complex patterns and how they might have been constructed. The quasi-crystal patterns in some of the later patterns in Uzbekistan indicate that a circle grid in its simplest form would not suffice for the construction of every pattern. Instead the underlying structure comprises of aperiodic tiling consistent with the artisan instructions given in the fifteenth century Topkapı scroll (Necipoğlu 1995). There is still an unexplored area to understand if and how the aperiodic layouts could have been introduced through circle grids.

Although an inquiry reaches beyond the scope of this text, it is worth to consider at least two issues to validate the circle grid. Firstly, the patterns Lu and Steinhardt look at, similar to those in the Topkapı scroll, are from a later era and are generally applied to ceramic tiles. Ceramic tiles easily serve as units of geometric tiling. On the other hand, when we look at the material evidence in the case of patterns applied on stone, as in the earlier Seljuk examples, physical tiles are rectangular blocks of stone independent of the geometry of the pattern. It is very well possible that there is a discrepancy in techniques and knowledge over time. Secondly, the circle grid gives the artisan flexibility to apply the pattern on uneven surfaces. Differently than with precut tiles, the drafting of a topology on an uneven surface allows for a continuous application.

In a practice based argument rather than a historical one, the application of the design on the material, in the case of stone, is a construction rather than tiling. This implies that it is a step-by-step process rather than an offsite prefabrication and on-site assembly of uniform tiles that is blind to seeing alternatives along the way. The circle based description of the process provides the basis for a richer computational understanding of the design of units in relation to the overall design of a panel. These patterns are then seen as a step-by-step construction that entails a lot of seeing on the designer-craftsman's behalf.

4   Variations on a Repetition

4.1   The Hexagon

The most distinguishable formal feature in the Seljuk patterns is the polygonal geometry. Equilateral polygons and corresponding stars are frequently used. In line with the cultural context briefly given in a previous section, the theological symbolism provides dominant formal features such as the star motif for the design from the beginning. The guilds guide the style and set restrictions on how much a designer could detach from the norm. On top of all this, the formal and technical aspects of the creative task introduce more physical constraints. The stone craftsmen work with simple tools   a compass, a straightedge, a chisel. Numeracy is not involved. Thinking in terms of these means and looking at end results, it is assumed that craftsmen initiate their patterns on guidelines drawn or carved lightly onto the material. For practical purposes—and to easily deal with bumpy surfaces, broken edges—these guidelines are drawn with the compass and were accordingly circles. In the examples where the repeated rule is *vesica piscis*, the width of the intersecting area is exactly one radius. Once the compass is at work again, with the radius at constant value, successive steps result in a uniform grid. Each intersection point becomes the center of a new circle. In the successive steps, the designer places the compass needles on the two points of choice. As the second circle is drawn intersecting with the first, new reference points emerge in addition to the center points. A circle grid topology emerges as seen in Fig. 3.3. This way the craftsmen can actually accommodate small mistakes that might agglomerate in their infinitely recurring pattern on irregular surfaces. All the constraints provide a context for cognitive and creative thought processes during design.

Commonly, the featured polygon is a hexagon. The equal sided hexagon pops up strongly once the equilateral triangle grid is superimposed on the circle grid. The designer most definitely knows and anticipates this symmetric and dominant polygon at the earlier stage of drawing the circle grid. Still, alternative polygons exist in other designs based on the same geometry. In one example, the design diverges to a non-uniform hexagon (Fig. 3.4).

Fig. 3.3 A step-by-step computation of a circle grid based on *vesica piscis*

Fig. 3.4 Two different designs on the same grid

The six points where a circle intersects with three axes of the triangular grid (twice on each) serve as references for the corners of the hexagon in the first alternative in Fig. 3.4. In the second, half of these references are changed to intersections of three neighboring circles with the three axes. The design breaks away from the equilateral triangle grid. A distorted hexagon emerges.

This switch to an alternative hexagon is significant in multiple ways. Firstly, the distorted hexagon is not a conventional fit to the circle grid. This means that despite the repetitive quality of the underlying system, the design vocabulary is not predetermined entirely. How and why the designer transforms the uniform layout to a distorted motif is surprising. Even if the style of the guild is imposing a structure, the designer acts on what he sees during design construction. The variety of patterns relies on such actions. These actions could be learnt actions. In other words, the designer may already have these as part of a vocabulary. This does not undermine their importance, however. These are visual jumps that change the course of a design whether premeditated or not. Secondly, some of the parts in the second alternative are kept the same as in the first alternative. Despite the acknowledgement of new intersections that can yield to all kinds of adventurous new lines, the designer repeats the hexagonal symmetries. In the end, the visual effect is just a little different while most parts repeat the relations in the first alternative. Thirdly, there is a founded relation between the two alternatives. Whereas each alternative seems to be arbitrary choices when viewed in isolation, their similar origins and features (parts) associate them with one another. We can appreciate this association only because we are observing both in the process and in comparison. This is the fundamental pedagogical tool in these pattern constructions.

What the designer sees is not significant from the point of view of this paper. Rather, it is significant that the designer sees. Knowing that valid alternatives are abundant, albeit dependent on purpose and context, I value these examples as illustrations of visual thinking at work. The dynamics are similar to those of ill-defined design problems where the problem definition keeps changing during design activity.

4.2   The Underlying Circles

As mentioned above, when the designer starts on the blank surface, the first step is the drawing of circles with the compass. Once the designer puts down the first circle, there is a new formal reference system to build on. The perimeter and the center point of the circle provide references to relate to. In the following stages, as more circles are added, what the designer sees (prefers to see) may keep changing, and in turn create new references. If the designer repeats the same rule, a consistent group of relations emerge.

The size of the first circle determines more or less the scale of the entire project in relation to the surface. Similarly, its location determines how the overall composition will be positioned. There is a good chance that they are related to the overall design in the craftsman's head: how the patterns will fit the edges of the surface or how the pattern is visible from a certain distance. It may well be that the possibilities considered by the designer are limited based on cognitive capacities, cultural restrictions, or contextual preferences. Nevertheless, it is useful to understand these choices as a part of a much larger set of infinitely many choices in order to appreciate the relations that are in fact implied in these choices. For example, *vesica piscis* is an instant on a series, given on the first line of Fig. 3.5, where the distance between the centers of two circles gradually increases. If we look at the spatial relationship of *vesica piscis*, two properties are easy to pick out: the radii of the two

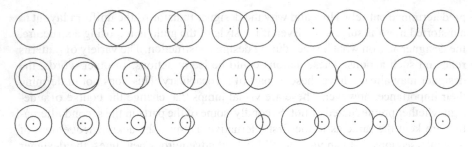

Fig. 3.5 Different spatial relations of two circles

circles are equal, and each of the centers is on the circumference of the other circle. If the second property is omitted, one can have indefinitely many spatial relations with two identical circles. What directs the choice of the designer is the parametric relation to the larger whole, which in this case can be the type of polygons anticipated in the next step or the final design. This helps in evaluation as well. And if the choice is to be reconsidered alternatives are known. If we keep increasing the distance between the centers but the radii of the two circles are not equal and have a ratio of 1:2 instead, a new series of possible relations emerges. Lines 2 and 3 in Fig. 3.5 illustrate spatial relations between circle couplets that vary in size.

All this speculation is assuming that the craftsman would be technically able to do this as well. With a compass of two chalks at the end of a rope, he can just fold the rope in half for the half measure. Some of these spatial relations with the half-size circles can guide him in dividing his initial circle, expanding or contracting it. The half measure is easy to guess. It is also relatively easier to attain for the craftsman. But he could also shorten the rope to arbitrary lengths by rolling it around the chalk or his finger. In many Seljuk patterns circles of different sizes result in motifs that combine varying sizes of stars (Bakirer 1992). It is possible to explore many other series where the ratio of circle radii is changing. All series are similar in principle. Most of their elements are equal in definition in terms of circumference and center relations. There are infinitely many series for all the different ratios of radii.

For the uniform circle grid that was given at the beginning of this section, the iterative application of a simple rule is sufficient. Nonetheless, if the designer leaves the constraint of using the intersection points as center for new circles, new results can be achieved. Still a regular grid, the one in Fig. 3.6 explores different spatial relations in addition to *vesica piscis*. This arrangement is also transformable to a tiling of decagons common in aperiodic examples and could be further explored to establish a link between circle grids and aperiodic patterns.

The designer's choices are in general very particular. He sets and follows rules repetitively but there are new reference points to consider each time rules are applied. In existing examples, there are more than a couple of ways to create polygonal infinity pattern guidelines with circles. Intertwined circles depending on how they are connected give different angled grids at intersection points. Executed examples are commonly based on a triangular/hexagonal grid, but rectangular/octagonal ones

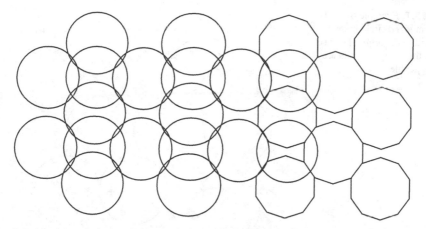

Fig. 3.6 An alternate circle grid that is similar to the arrangement of common decagonal tiles

are also frequent. Whereas the hexagonal comes out so naturally in a process with the compass, the octagonal one is possible as long as the designer is using a parallel ruler (Fig. 3.2). Some of the most stunning examples are pentagonal. One could do much more with the same two circles. The design space is infinite.

4.3   Size of Parts

The underlying system of repeating circles sustain properties such as axial and rotational symmetry, closed or incomplete polygons, and lines that extend over the surface. These are repeated features that help us identify the style and are dependent on the circle grid. Deviations in pattern designs mostly start when straight lines are superimposed on top of the circles. The design space is infinitely rich during the design process and the designer's path is deemed finite and unique only after the fact.

Figure 3.7 shows another example of a possible variation. During the design, once a circle grid is created, one can also invent fields of different densities. The designer may see new points of reference for himself at the intersections of circles and straight lines. New circles of a different size may emerge from this set. Instead of lines then, one could be drawing circles to connect the intersection points. Similar to the possibilities at the very beginning of the design when the canvas was still blank, new circle grids can be acquired. In the same circle grid, an alternate grid of larger circles is constructed and areas with larger hexagons emerge (Fig. 3.7).

There are indefinitely many descriptions to a developing shape like this. There are just as many descriptions for connecting the lines in the intersection point grid. Recognizing new parts in a shape is possible only if the world is viewed in part relations rather than distinct and preset object properties and classifications. Designer's purposive actions determine the choices along the way and the final results. The

Fig. 3.7 Example of a different sized circle juxtaposed to the original one so that the sizes of motifs are changeable

alternatives to a designer's path may not be legitimate without a real context, but the process is ideally open to just about anything.

4.4 *The Weave*

Once the shape is finalized and the craftsman starts using his chisel to carve, a new level of materiality is introduced. The craftsman calibrates how thick he wants his shapes to be according to the brittleness of the stone and the space in between the shapes. He then proceeds to carve out the centers. The braiding comes later as a "face treatment". Nevertheless, just as much as the earlier process that takes the design from intersection points to hexagon, the weaving effect suggests that the craftsman's view is holistic. The weave pattern repeats in most examples. There is a pretext: the designer establishes which end of the line he starts hitting with the

Fig. 3.8 The alternating weave (from *left* to *right*)

chisel or what sequence he follows. In the overall composition, the order of which line(s) gets chiseled out first is important. Variations in that order would result in hexagons that interlock differently as seen in Fig. 3.8.

In other motifs, we see that the designer is sometimes able to depict infinite rays rather than enclosed shapes, despite the hexagonal or octagonal geometry of the overall design. This is especially impressive and confusing if the final motif alludes to polygons but the lines acting out as design elements do not follow a predictable weave pattern. The design can consist of rays that shoot off the edges of the frame instead of interlocking polygons as in Fig. 3.9 which shows a reconstruction from an actual motif in Kayseri.

5   Seeing the Broader Picture

There are endless possibilities for each of the four variation themes above. There are also many more themes one can explore. The design strength of the Seljuk patterns comes from the abundant potential to explore a versatile structural canvas, namely the interlocking circles. There are surely other examples in design to argue for the value of rhythm as a relation of repetition and variation. At a first glance, geometric patterns may seem to involve less creativity in comparison to the complex formal exercises such as those in architectural design. However, these patterns have the essential minimum to talk about visual richness and the variety in perceiving parts. The restrictions of the trade make it easier to pinpoint to that minimum. The visual jumps that change the developmental course of a pattern design are moments of selection in an infinite space of possibilities. These selections are not predictable to those outside the trade. Still, with the right balance of repetition and variety, meaningful designs are possible.

Fig. 3.9 An existing pattern with a complex weave and rays that extend beyond the frame

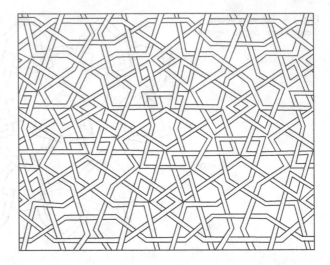

Meaning in both cognitive and creative processes is associations one makes with percepts, through recognized features—regularities—in specific contexts. Richards et al. (1992) prompt the idea that key features in objects and events play an essential role in acquiring meaning in perceptions. The argument follows that in an effort of constructing meaning in an object, preference is usually given to the most familiar/ associated feature in that context. This is all framed by the regularities one has so far acquired. Preferences, and thus regularities are context specific. They are formed as part of our formation, and pertain to a particular world. People with different formations, thus different contexts, have different preference orders for features. Design process is rarely the conversion to a key feature that is preset. Rather it is the diversion from it. This holds true for variations among the repetitive features of Seljuk patterns as well. Designers rely on repetitions to diverge from in pursuit of novelty.

In general, a designer's process shows that one specific individual's perceptive context can vary from time to time. This is crucial in a creative process. One must acknowledge that a key feature is only a transiently ideal condition, even for one individual. The properties can be reinterpreted and new ideals can be devised to address different intentions. The key feature is the result of the Gestalt perceived at that present time. This is possibly why the craftsman is not following the grid *per se*. His decisions are based on other key features than those assumed by us, outsiders, or even what he initially thought they were. The social, material, historical contexts all impose on perception. Discoveries are made in the canvas as the eye switches from one coordinate frame to another. Differences to be discovered among repeating features are just as much a positive challenge to the designer as they are to the viewer.

The ways in which this paper looks at the designer's process leaves out a lot of the trivial conditions. The designer or the craftsman must have hundreds of different seeing instances on an irregular stone surface, with his chisel tapping in the wrong direction once in a while, or tapping a tad too much into some of the shapes. Also, he is probably more spontaneous than in the limited illustrations of the process I show here. If we are to look at the original design, most of the shapes are not exact. There is even more room for variation through the imperfections. The geometry that the craftsman uses gives way to indefinitely many perceptions and designs. There are about 100 or more designs alike that are known. All this spontaneous action operates on changing key features and preferences.

Seljuk patterns are full of repetitions, mainly because of the craft bounded by technical and cultural traditions. However, within all the limits, there are points of freedom, especially if we accept the production of these patterns as a visual process where the designer takes the initiative. Hence, all part relations cannot be defined beforehand. They will arrive at the scene just as the designer does, but they cannot be predicted. It is not only inefficient but simply inhibitive to foresee that definitively. If I am to define absolute features for a circle, which polygons will I be missing out on in later steps? Circles are the beginning guidelines only and they set a consistent ground to walk on and break where necessary.

Choosing from options is an incomplete description for design which is really a decision process in an act of creation. A designer assumes that his/her choices are close to being unlimited each time there is a decision to be made. The choices are unlimited despite a rationale that is deliberate and particular to the designer and to the circumstances. The designer, at any point, can shift his/her reasoning to another consistent path of thought. The visual step-by-step computation of the patterns, as assumed in this paper, undermines presumptions that will inhibit the process at any stage. It is essential that these patterns are visually constructed and not blindly copied entirely based on a previously configured model. Visually informed decisions can reflect a unique path of the designer's preferences. This is a lesson to be learnt for modern day—design—computing.

From the design point of view, approaching these patterns as computation is significant because *the relation of that unique path to the broader design space is better appreciated when formal relations between design elements are analyzed comparatively.* In addition to better understanding the working methods of the period, this approach, with potential to turn into a design learning methodology, may be used to understand and convey how to create good designs.

Acknowledgements All illustrations are by the author. Thanks are due to Whitman Richards, George Stiny, Sibel Tari and Nyssim Lefford for stimulating conversations on the matter.

References

Abas, S. J. and Salman, A. S.: 1992, Geometric and group-theoretic methods for computer graphics studies of Islamic symmetric patterns, Computer Graphics Forum 11, 1, 43–53.
Bakirer, Ö.: 1981, Selçuklu Öncesi ve Selçuklu Dönemi Anadolu Mimarisinde Tugla Kullanimi {The use of brick in Anatolian architecture in pre-Seljuk and Seljuk era}. Ankara, Turkey, ODTÜ.

Bakirer, Ö.: 1992, ' "Aksaray Cincikli Mescid'in On Yuzundeki Geometrik Orgu Duzenlemelerinin Tasarimi Icin Bir Deneme" ' IX. Vakif Haftasi Kitabi: Turk Vakif Medeniyetinde Hz. Mevlana ve Mevlevihanelerin Yeri ve Vakif Eserlerinde Yer Alan Turk-Islam Sanatlari Seminerleri, 2–4 Aralik 1991 Ankara (Vakiflar Genel Mudurlugu Yayinlari, Ankara, Turkey).

Bakirer, Ö.: 1999, Story of the Three Graffiti, Muqarnas, 16 pp. 42–69.

Critchlow, K.: 1976, Islamic Patterns: An Analytical and Cosmological Approach, London: Thames and Hudson.

Feldman, J.: 1992, Constructing Perceptual Categories, Proc Comp Vis and Pat Recog, Champaign, IL, June, pp. 244–250.

Grünbaum, B. and Shephard, G. C.: 1987, Tilings and Patterns. W. H. Freeman.

Grünbaum, B. and Shephard, G. C.: 1992, Interlace Patterns in Islamic and Moorish Art, Leonardo, Vol. 25, No. 3/4, pp. 331–339.

Kaplan, C. S.: 2000, "Computer Generated Islamic Star Patterns" in Bridges 2000: Mathematical Connections in Art, Music and Science.

Lawson, B.: 2001, Language of Space, Oxford: Architectural Press.

Lee, A. J.: 1995, Islamic Star Patterns, Muqarnas, Vol. 4, pp. 182–197.

Lefebvre, H.: 2004, Rhythmanalysis: space, time, and everyday life, Trans. S. Elden and G. Moore, New York, NY: Continuum.

Lu, P. & Steinhardt, P. J.: 2007, Decagonal and Quasi-Crystalline Tilings in Medieval Islamic Architecture, Science 23: 315(5815) pp. 1106–1110.

Mülayim, S.: 1982 Anadolu Turk Mimarisinde Geometrik Suslemeler (Kultur ve Turizm Bakanligi Yayinlari, Ankara, Turkey).

Necipoğlu, G.: 1995, The Topkapi Scroll-Geometry and Ornament in Islamic Architecture: Topkapi Palace Museum Library MS H. 1956, Santa Monica: The Getty Center for the. History of Art and the Humanities.

Ogel, S.: 1966 Anadolu Selcuklularinin Tas Tezyinati (TTK, Ankara, Turkey).

Ozdural, A.: 2000, Mathematics and Arts: Connections between Theory and Practice in the Medieval Islamic World, Historia Mathematica 27, pp. 171–201.

Özkar, M. & Lefford, N.: 2006, Modal relationships as stylistic features: Examples from Seljuk and Celtic patterns. JASIST 57(11) pp. 1551–1560.

Rasmussen, S. E.: 1959, Experiencing Architecture, Cambridge, MA: The MIT Press.

Richards, W., Feldman, J. and Jepson, A.: 1992, "From Features to Perceptual Categories" British Machine Vision Conference 1992 eds. D. Hogg & R. Boyle. (Springer-Verlag, New York).

Shlokovsky, B.: [1925] 1991, "Art as Device," translated by Benjamin Sher in The Theory of Prose, Bloomington, IL: Dalkey Archive Press.

Stiny, G.: 2006, Shape, Cambridge, MA: The MIT Press.

Chapter 4
Brain, Technology and Creativity. BrainArt: A BCI-Based Entertainment Tool to Enact Creativity and Create Drawing from Cerebral Rhythms

Raffaella Folgieri, Claudio Lucchiari, Marco Granato and Daniele Grechi

We developed BrainArt, which is essentially a workbench that allows users to create drawing using their own cerebral rhythms, that are then collected by a commercial BCI device. The application gave us the possibility to observe and analyze the spontaneous creative expression of the participants' insight during an experimental session, where users created an artwork using the application. Therefore, this work presents both the BrainArt application possibility for entertainment and freedom to express individual conscious and unconscious creative insight, as well as the preliminary experimental results obtained by a combined approach of statistical, signal and behavioral analysis.

1 Introduction

In this paper. we present BrainArt, an entertainment application allowing users to create drawing through the interpretation of their cerebral rhythms. For the interaction with the BrainArt workbench (better described in the related paragraph) we chose to use EEG-based BCI (Brain-Computer Interface). In fact, the progress in Brain Imaging techniques gives strong impulse in observing the brain in action (the so called "living brain"), and, thanks to MEG, fMRI and EEG, researchers can study the response to specific stimuli in real time during specific experiment. Due to its lower cost, if compared with other methodologies, and especially thanks to high

R. Folgieri (✉)
Department of Economics, Management and quantitative Methods,
Università degli Studi di Milano, Milan, Italy
e-mail: raffaella.folgieri@unimi.it

C. Lucchiari · M. Granato
Department of Health Sciences, Università degli Studi di Milano, Milan, Italy

D. Grechi
Department of Naval, Electrical, Electronic and Telecommunications Engineering, Polytechnical School, Università degli Studi di Genova, Genoa, Italy

N. Lee (ed.), *Digital Da Vinci,* DOI 10.1007/978-1-4939-0965-0_4,
© Springer Science+Business Media New York 2014

time resolution, EEG represents a promising approach in investigating individuals' reaction to stimuli in real time. BCIs (Brain Computer Interfaces) devices, derived from the EEG medical equipment are now extensively used in the entertainment and in the scientific communities due to their reliability in collecting EEG data with a high time resolution, and to their portability and low-cost. BCIs are largely applied in entertainment applications, allowing users to interact with game/educational environments through their cerebral rhythms. For these reasons, and for their easy usage, they appear to be efficient and innovative interaction devices that enable users to not only interact with different virtual environment, but to also express their conscious and unconscious creative insight.

Thanks to the possibility of registering users' brainwaves (for a later playback of the session) given by BrainArt, we also performed a study on creative high-level mechanisms of individuals. The general aim of this study consisted in analyzing the correlates and the precursors of the creative act. The study considers an opposite position with respect to the traditional one (neuroimaging): in fact we do not "simply" study the neuropsychology of the creative act, but instead we want to give the participants in the experiment the opportunity to express their own creativity in an unconscious way. Therefore, they are able to express what is intimate by nature, and so, are impressive more than expressive. Only in a second moment, will the individual express his/her own aesthetic judgment, choosing the image that is closer to his/her idea. In this phase, conscious and unconscious processes meet each other. Indeed, we developed BrainArt within a cognitive framework where creativity and art expression are conceptualized as the result of a dynamic between conscious and unconscious processes, and yes, creativity may be considered a borderline state of mind, in which ideas and images fluctuate in a near-consciousness state. When the idea arises to the consciousness, the mind turns back to its previous equilibrium, and divergent thinking is replaced by canonical thinking. Thus, the mind functioning may be described as a series of temporary equilibrium, or balancing between implicit/automatic and explicit/controlled mechanisms blended in such a way to respond to cognitive and behavioral demands. For instance, when we are naïf at a given task, we usually use conscious and controlled processes since we need to consciously direct our attention to each step to be performed. This process requires a lot of resources, and furthermore, it requires that we already know what to do in each single step (e.g. to park a car in tight parking site). When we get skilled at the opposite, we rely on unconscious and automatic processes, letting things happen in a fast and accurate manner.

However, when we face a puzzled task and we need to explore landscapes of possibilities, we need to find a different balance between conscious and unconscious. This balance is what we can call creativity: a sort of dedicated state in which the brain gets focused on an idea or a task in order to follow solution paths that our conscious mind would never imagine. Indeed, when we navigate the world using the consciousness we need to find rational reasons to take decisions; we need to find a reassuring anchor in order to keep anxiety and arousal at adequate levels. In other words, consciousness navigation limits us within the bounds of self-perceived safe and well-known paths. However, humans conquered the world overcoming boundaries, and what sets us apart from computers, is that our mind has the ability to find innovative solutions, instead of remaining stuck in old cognitive and behavioral patterns.

We can hence say that the possibility of finding different cognitive balances in different contexts is probably the secret of the "success" of the human mind, in regulating human behavior as compared to animals. At various levels, in fact, in all vertebrates, is it possible to observe a certain dynamic between these different processes. In higher primates we can identify a true equilibrium, in which conscious processes can adjust to unconscious ones (and not just vice versa). The creative act or the insight is borderline phenomena, in which conscious and unconscious levels are mixed.

Starting by this framework, we designed and developed an entertainment application in which creativity is enacted by unconscious process, but in collaboration with conscious motivation. Our aim was to activate a "creativity-dedicated" balance between conscious and unconscious processes in order to obtain a state of mind similar to the spontaneous creative process, but directly guided by brain activity without the intervention of verbal and semantic modulations.

In this perspective, we also implemented a "simple" experiment, even if explorative and absolutely preliminary, which can be considered paradigmatic. Indeed, conscious and unconscious processes are explicitly evoked, but in such a way that the final result cannot be considered either a conscious or an unconscious product, but instead, a blend of both. In particular, the study of the EEG frequency bands allows us to analyze the unconscious dynamics that precede creative acts.

On the basis of the following discussion of the results, can we claim that a theta or a beta/theta threshold exists, and whether it is capable of defining the transition from the impression to the expression? It seems that the beta/theta relationship draws significant results only when the individual has an artistic education. May this mean that when the subject is more aware of his/her own creative processes (having developed specific skills or knowledge), will the beta rhythm report an imbalance or "rebalancing" toward consciousness that is toward conscious processes?

This work represents an opportunity to underline the importance of gaming also in a Cognitive Science perspective. In fact, Cognitive Science studies the mechanisms of natural and synthetic intelligence, so, a game environment, for its nature, gives the possibility to observe them in their spontaneous expression. This approach is particularly suited to study the relationships among Art and Brain, as illustrated in the next few paragraphs.

The paper is organized as follows. In Chap. 2, we introduce some basic concepts of psychology, of art, and the neuropsychology of creativity. Then we discuss the relationships among Art, Brain and Technology, in order to establish the framework that led our work in building the BrainArt application. At the end of the chapter we focus on the role of Technology in Art, for the preservation and new expressive opportunities offered, and for the possible scientific opportunities that tools like BrainArt offer in the enhancement in our knowledge of brain mechanisms and the development of new cognitive neurocognitive paradigms.

In Chap. 3 we present the BrainArt workbench with regard to the application itself and the experiment performed. Finally, in Chap. 4, we summarize our conclusion and present further possible developments.

2   Art, Brain and Technology

Studying Art both from the point of view of the creative act, and from the point of view of observers of masterpieces, allows comprehensive investigation of the processes which underlie the interaction between Brain and environment, exploring the connections among the cognitive, creative, interpretive and expressive processes. This is the core interest of Cognitive Science in studying the connection between Art and Brain, since the discipline of Cognitive Science is aimed at understanding the mental processes underlying the different acquired or innate abilities and skills. This definition highlights the vastness and the complexity of the investigation field, and consequently, the need of a multidisciplinary approach as suggested by Norman (Norman 1981), who identified several fundamental disciplines. For the purpose of this study we will refer to Artificial Intelligence (A.I.), Psychology, Neuroscience and Neuroaesthetics.

The interest in studying the relationships between brain and art consists in detecting the mechanisms of creativity, to realize models able to simulate intelligent systems or to be used by them. Understanding creativity mechanisms is the key to discover how people move from skilled performance to problem solving, how they interpret visual or audio stimuli, and communicate. Modern Cognitivism is, in fact, also identified as "Human Information Processing", because it investigates on humans' approaches to process information.

Exploring the links among Art, Brain and Technology entail investigating the relationship between Psychology and Art, Neuroscience and Art (Neuroaesthetics), and among Technology, Art and Neuroscience, as follows in the next paragraphs. Among the Cognitive Science disciplines, Artificial Intelligence (A.I.), jointly with the innovative technology at disposition today, provides not only a means to better involve the public observing Art, but also allows one to explore the creation of artistic expressions and, at the same time, to analyze the processes behind creativity, the artistic thought and the effect of the stimuli offered to observers by Art.

2.1   Psychology of Creativity. Eureka (I Have Found it!): Creativity as an Interplay Between Conscious and Unconscious Processes

The creative process moves through stages, beginning with the preparation stage at which the basic information is collected and skills are developed. This is followed by incubation: a relaxed phase during which the person does not use the conscious function to solve a problem, but some connections arise unconsciously. This then leads to inspiration: the so-called eureka experience when the person suddenly "sees" the solution. It ends with the production: a moment in which insights are expressed in a useful form. The specifications of this basic process will vary according to the type of creativity, e.g. writing a novel is different from identifying a new chemical synthesis, but the basic process and the principles seem to be the same in the different types of creativity.

Describing their subjective experiences, creative people often say the same things over again:

"I cannot force inspiration. Ideas come when I'm not looking for them, when I'm swimming, or running, or am standing in the shower." "It happens as if by magic." "I can see things that others cannot, and I do not know why." "The muse is sitting on my shoulder." "If I focus on finding the answer never comes, but if I let my mind simply wander, the answer is inside of me". (Andreasen 2005)

If we try to understand these descriptions using the framework provided by our understanding of the mind and brain, then we are led to conclude that the creative process stems from a change in the balance between conscious and unconscious processes. The person seems to enter into a kind of "pro-creative" state in which the mind wanders thoughts and images freely floating, without censorship. During this "fluid" phase, the brain is probably working feverishly, in spite of the subjective sense of relaxation. Indeed, neuroimaging studies showed that a free associations task implies an enhancement of the interplay between associative cortices, that seem to communicate each other in a free and uncensored way. When subjects are questioned about their mental activity during this condition, they generally describe it to be one where they are engaged in a free, random, and self-referential flow of thoughts, characterized by the activation of episodic memory contents. This state of mind is generally called "flow" during which what we know is somehow connected with the unknown in a focused and highly concentration status (Csíkszentmihályi 1990). In this concentrated state, a reduced activity in the prefrontal cortex is generally found, suggesting that decision making and working memory processes are somehow displaced by creative processes. Brain imaging studies showed that this executive and high cognitive function silencing is coupled by the activation of the basal ganglia, and is able to modulate implicit, and then unconscious, cognitive functions.

In a general perspective, creativity is the result of evolutionary plasticity of the human brain that allows adaptation to the environmental demands through new strategies based on experiences. In humans, the collaboration of the basal ganglia system and the frontal cortex network lead to a growing separation between implicit cognitive functions (skills and procedures unconsciously available to perform a task) and explicit processing systems (based on rules, verbalizations consciously accessible). The implicit system is wired in basal ganglia and the cerebellum, while the explicit system requires a complex network made by the prefrontal and medial temporal cortexes. However, these two large circuits interact each, particularly through the *striatum* structures. The other giving rise to a dynamic balance between explicit and implicit processing. Different balances lead to a different state of mind and, consequently, to a different way of experiencing and reacting to the environmental stimulations.

During a pro-creativity state, the striatum and the prefrontal cortex work in symbiosis for the generation and selection of new ideas. In particular, according to Oliverio (Oliverio 2009), the stratum acts as a kaleidoscope, and projects memory sequences, procedures, and emotional evaluations onto the prefrontal cortexes that in turn will select the items, hence inhibiting irrelevant associations and facilitating the attention-focusing in order to solve a problem (Oliverio 2009).

Beethoven thus described the process of creative production:

> I carry around an idea for a long time, often for a very long time before writing it down on a paper. On the other hand I am sure that I will not forget it, even after many years. Change a lot, waste a few things, I try and try and try again until I am satisfied then I start to develop the work in my head, the basic idea never leaves me. It takes shape, grows: I can hear and see the images in front of me, from every angle.

2.2   Neurosciences of Art: The Neuroaesthetics

In recent years, we have made great strides in understanding the neural basis of the subjective experience and cognitive processes of the human mind. The brain structure plays an important role in constraining the expression and perception of the subjects. Similarly, the brain architecture sets limits and at the same time, opens up possibilities to artistic freedom and creativity. Neuroaesthetics is the discipline whose goal is to find the neural basis of mental processes precisely related to art. It is revolutionary in part, and in part simply the extension of previous work. The term "aesthetics" is not recent and it tends to coincide with the critical reflection or philosophical about the categories that are related to art (Baumgarten 1735). From the end of the XIX century, the discussion on aesthetics has been based on a new plan that integrates the scientific knowledge on perceptual and cognitive processes with the traditional philosophical discussion on some priori categories inside which moves the art. The scientific study of mental processes has been accompanied by a series of attempts to tie the artistic experience in these processes. Above all, there has been a more consistent and methodical convergence of interests between the disciplines of psychology and art criticism, even as a result of input from the avant-garde and post avant-garde. Many studies are specifically dedicated to the relationship between art and psychology, linking a certain artistic phenomena through the links with human perception. The work of art is no longer viewed in isolation as an object to, but as a combination of materials, shapes and perceptions. The search for regularities and principles that underlie the perception of an artwork by and the observer is the main goal of neuroaesthetics, in which it is possible to find two basic guidelines. The first investigates mainly the scope of the vision: the conditions in which we see preferences and criteria ordering the visual world. The second is aimed instead to analyze the relationship between viewer who perceive and the world that is perceived, since the art holds a privileged position in this relationship.

Indeed, artwork makes us see more than "natural" things evoking idiosyncratic aesthetic experiences well beyond the given information.

The work of Semir Zeki, the neurologist who congaed of the term neuroaesthetics in 1999, must be considered fundamental to understand the meaning of the discipline. He argues that color vision is interesting as an object of study because understanding how the brain constructs colors allows us to investigate the brain processes that underlie the aesthetics in the sense of the ancient Greeks, that is the acquiring of knowledge through the senses (Zeki 2004). It is important to recognize that sensitivity opens the subject to the world, through which you can reach the intellectual knowledge, but is also in itself a bearer of knowledge, limiting our experience when we accept the bounds of the natural world.

Aage Brandt (Brandt 2006) proposed to consider neuroaesthetics as the study of neuronal processes of perception and mental organization of cognitive activity stimulated by the artwork, following both a cognitive and evolutionary approach. Evolutionary process and neuronal process however, are not the same. Olaf Breidbach (Breidbach 2003) distinguishes between an evolutionary aesthetic and a neuronal aesthetics. The aesthetic lies not merely in a sign, but in his interpretation of the brain and then the objective is defined with a construct produced by the subject (Breidbach 2003).

Focusing on the content of the painting as an expression of the ability of a visual artist and the observer, removes any distinction, or even any collaboration between form and function, content and style, shape and technique. To its extreme consequences, the artist is a neurologist in action, who not only knows the laws of the brain but also knows how to put them into practice. Furthermore, since the brain changes over time and responds plastically to stimuli, then the development of his artistic production must be predictable. However, such predictability is in doubt because the organic predisposition to be stimulated by sensations, on the one hand, and the actual experience, secondly, is heterogeneous. Zeki (Zeki 1999) gives other three reasons to be cautious about predictability of artistic production: the first is the role of the unconscious and its power; the second is the assumption that the physiological stimulation of specific visual areas can create perceptual experiences, but the aesthetic experience depends on the whole brain activity; and the third reason is that the image is completely emancipated from its object: basically a picture cannot represent an object, since the brain can simply imitate the object in some particular aspects (Zeki 1999). The aesthetic experience of a work of art integrates and transforms the individual perception of reality in a lived experience with regard to the subject: the artwork disturbs him, excites him, soothes him and the like (Geiger 1928). According to Zeki, what is possible by virtue of situational constancy, which focuses the attention of the observer, by introducing it in its memory to retrieve similar past events.

An idea can be described in reference to the functions and the actual functioning of the brain, since it is the representation stored in the brain of the essential characteristics seen and selected. Neurologically, it's not possible to give rise to perfect visual forms without being exposed to the visual world practice. The task of brain function would, therefore, precisely represent objects as they are in reality

(Zeki 1999). In the sensory-motor system there are neural structures that are secondary, in the sense that they are neurally connected to the system immediately involved in perception and movement. For example, the premotor cortex enacts complex actions through the connection with the primary motor cortex, which controls simple actions. When this connection between premotor cortex and the motor cortex is inhibited, the secondary premotor cortex can function as the ring of a transmission chain, a cog, which can still compute complex configurations that allow us to carry out inferences and that may change over time. These secondary configurations, so-called cogs, give rise to what we perceive as an artwork. Till now, many types of cogs have been proposed, and each type corresponds to a formal aspect (Gallese and Lakoff 2005). The cogs, which are embodied at the same time because they are part of the sensory-motor system and abstract since it does not include any details, allow us to have an immediate understanding of the embodied form of abstract art, and metaphors applied to structures cog provide interpretations of abstract art. Lakoff makes use of the notion of schema assumed to be an interactive network that depends on the nature of our body, our brain and the social and physical interactions with the world (Gallese and Lakoff 2005). The richness of the notion of schema for neuroaesthetics was also confirmed by Stafford in another sense. A schema is a recursive and fluid configuration, which maintains its internal relationships of correspondence from time to time. The worth of this concept is to hold together biology and culture to reconstruct the gap between form and function: the set of points, lines, surfaces and circles, triangles and squares that govern the formation of the planet and the attempts of survival are intuitively understandable because the human brain has developed the same form (Stafford 2005).

The scheme is therefore the grammar that regulates thought and concrete experiences. Gombrich (Gombrich 1959), pointed out how the man is going into the pictorial construction of the world through the rhythm between the use of a model and the corrections that are applied, so that the conquest of naturalism can be defined as the gradual accumulation of the corrections suggested by the observation of the natural world.

Mirror Neurons and Creativity: The Brain as Simulation Tool

Mirror neurons are bimodal nerve cells activated when we perform a certain action, but also when we observe others performing the same action particularly among animals of the same species. Indeed, seeing is not just a passive behavior, but it already involves a sort of unconscious mental simulation of the environment we are experiencing. By the activation of the mirrors system, the observation of others' behavior activates our brain as if we were doing the action ourselves. These neurons were discovered in 1987 by Giacomo Rizzolatti (Rizzolatti et al. 1997), while he was working with Leonardo Fogassi and Vittorio Gallese at the University of Parma.

According to Stafford (Stafford 2007), the discovery of mirror neurons, considered as a sort of wireless communication tool, takes the mimesis concept

back in the centre of the aesthetic debate. In particular, the mirror system functioning raises the question of the relationship between first-person and third-person
experience in the construction of knowledge. If, on the one hand, as indicated by
Berenson (Berenson 1948), when I immerse myself in imaginary, I feel my life
intensified, on the other hand, the conception of a mirroring and simulating brain
mirror opens to a kaleidoscopic notion of identity. According to the latter notion, the
identity of a subject is built on a series of narratives based on first- as well as third-
person experience. That same notion may be extended to include the relationship
between conscious and unconscious processes. Indeed, we are not aware of most
processes and experiences through which we construct our identity, but conscious
and unconscious processing ought to find a balance in order to sustain an adequate
control of our experience. When this balance is broken, a pathological shift in conscious state is experienced. For example, people who suffer from the Capgras syndrome are convinced they do not see themselves in the mirror, but someone else,
or they feel that a well-known face is actually hiding an impostor or an alien who
took the place of the known person. In this case, the brain area of face recognition
is disconnected from the limbic system that unconsciously processes emotions, so
patients recognize the identity borne by a face, but they don't feel to recognize the
person (something is then felt wrong, in their experience).

Even though these processes become "clear" only in pathological conditions,
the experience of art offers us the possibility to investigate similar mechanisms in
healthy people as well. For instance, Worringer (Worringer 1907) has introduced
the distinction between an impulse for empathy, for which we enjoy ourselves experiencing an artwork, because we project onto a line and in a form our sense of life,
and a pulse of abstraction, which aims to isolate a subject from its natural context.
Both impulses respond to a need for self-alienation, for which we get lost, falling
empathically in an artwork, or separating ourselves through an abstraction process
to elude uncontrollable experiences. The first type of individual alienation, collapse, can lead to a total loss of self and to a crisis, known by the name of Stendhal
syndrome, which may be of clinical interest. This relationship between empathy
and abstraction in the subject produces a not illusory identification, which is one of
the preconditions of aesthetic enjoyment, for which the immediate viewer experience is suspended, and a real or a hallucinatory experience materializes.

Indeed, art has the task of producing an image or a simulation of the reality,
albeit without obliterating or removing the observer personality and knowledge of
the natural world. Art, then, does not give rise to an illusionary world, but it extends
mind possibilities, opening new frontiers and unblocking landscapes and options,
thanks to an "as if" experience (Pinotti 2005). This "as if" experience doesn't only
imply a falsehood world, but it also allows the viewer to believe without deception,
to see and hear as if he/she is acting on his/her own. Mirror neurons therefore lead
to reflection on at least the first stage of knowledge construction, as when a person
is put in front of an artwork. This conception of the "as if" brain can be integrated
with what is stated by Changeux (Changeux 1994), for whom the contemplation of
the viewer is a recreational activity, and in this re-creation, he goes so far as to at-
tribute mental states, emotions, intentions to characters of the artistic composition,
thus mentally tracing the path of the artist's work.

Similarly, Freedberg and Gallese (Freedberg and Gallese 2007) observed that in front of a photograph such as of a dead man, our brain responds as if that picture convey first-person information, and for a moment one is left with a slight feeling of anxiety and despair. The physical involvement with an image results very quickly in some emotion. Hence, the aesthetic response is the result of the activation of a built-in mechanism enacting the enjoyment of an artwork. These mechanisms are activated not only in front of iconic or figurative works but also by abstract works as Pollock's drip paintings, for which the audience frequently experiences a sense of involvement with body movements that are implied by physical traces left by the creative actions of the painter. Two are, in Freebberg and Gallese view, the relationships involved in the use of pictorial works: that first is between the empathetic feelings embedded in the observer and the representational content of the work in terms of actions, intentions, objects, emotions and sensations painted, thus leaving aside the thematic content of the picture. The second relationship relates to the motor gestures with which the artist created the work and what they remind to the user: signs, prints on canvas or material, are the visible traces of movements directed to obtain the final result; they are therefore able to activate the corresponding motor area in the brain of the observer (Freedberg and Gallese 2007). Hence, no aesthetic judgment is possible ignoring the simulation work of the mirror system: to achieve the wanted result the artist ought to use, consciously or not, the knowledge of the body to evoke empathy and involvement. This concept can help to understand that the tension between abstraction and figuration is not only about the degree of similarity between perceptual-iconic art and the natural world, but about the translation of the outward, visible, figurative representation in a virtual world to be simulated in the viewer's mind.

2.3 Technology, Art and Neuroscience

The possible connections between Art and Technology, related to creativity and to the study of its mechanisms, are mainly three:

- the opportunities offered by technology in art and cultural heritage preservation;
- the potentialities provided by new tools and devices for the artistic expression;
- the possibility to study the links among Art, Brain and Technology and, jointly, the creative processes, allowed by the progress of A.I., by Brain Imaging and technological devices.

Art and Technology: Representation, Preservation, New Opportunities

New technologies, such as A.I. models for the so-called "big data" (a large amount of visual, textual, audio data) processing and tools and applications like Virtual Reality, Internet and Augmented Reality, represent a great opportunity for art, both in order to preserve and to bring the public closer to it, extending contents and making them available to a mass public.

Today, protecting cultural heritage against the process of deterioration represents a great challenge, and most agree on the need to preserve either tangible or intangible works for the benefit of future generations. For example, Virtual Reality plays a fundamental role not only in contributing to the preservation of art, but also for the fruition of the art masterpieces in a new way. In the specific field of cultural heritage, 3D representations and Virtual Reality have been applied especially in reproducing ancient scenarios, artefacts or monuments currently in a deteriorated state. In cultural heritage, and specifically in museums and visitor centres applications, usually the immersion characteristic is not implemented, due to the difficulties of involving a large number of users through expensive technologies. Nevertheless, there are many projects in which Virtual Reality, and more generally, the 3D reproduction of products and environments, is applied to involve the public presenting the paradigm of "walk-through" for a virtual visit. One of the first heritage application of this kind, in 1994, was the virtual reconstruction of Dudley Castle in England, showed by the British Museum in a conference held in November 1994 (described in the technical paper *"Imaging the Past—Electronic Imaging and Computer Graphics in Museums and Archaeology"*). Other remarkable works include an interactive virtual journey inside Leonardo da Vinci's, Michelangelo's, Alberti's, Botticelli's and Brunelleschi's masterpieces, and *CityCluster*, an online VR network platform connecting people from remote locations[1].

Recently, a low-interaction kind of Virtual Representation exists in the possibility given by Internet and mobile devices, to navigate within museums and visitor centres in a 3D environment reproducing the real site. Significant Interactive examples of this application are the virtual reproduction of the *Cappella degli Scrovegni* (Padova 1303–1305, Giotto from Bondone)[2], or the virtual tour of the *Cappella Sistina* (1475–1481, Michelangelo Buonarroti)[3], in the Vatican State. We also need to cite the *Google Art Project*[4], aiming at collecting masterpieces and navigable rooms reproduction from museum all around the world. Other examples of virtual representations are the famous *Villa of Livia*[5], the reconstruction of ancient Egypt sites or even those about Ancient Rome realized by Altair4.

Several Museums and Cultural Centres are experimenting the adoption of new technology devices to bypass linguistic, cultural and geo-localization barriers. This fact is evident if we think of the possibilities offered by Q codes and by Internet links to obtain additional information (Bait et al. 2005). The potentialities of the use of visual devices in an Augmented Reality perspective, for example, represent an interesting possible improvement in the fruition and in the emotional and cultural exploration of Art (Banzi and Folgieri 2012; Folgieri 2011). Augmented Reality

---

[1] http://www.fabricat.com/FISCHNALLER_MAV_workdescription.htm.

[2] http://www.haltadefinizione.com/magnifier.jsp?idopera=15&pagina=1#multi&lingua=it.

[3] http://www.vatican.va/various/cappelle/sistina_vr/index.html.

[4] http://www.googleartproject.com/.

[5] http://www.vhlab.itabc.cnr.it/flaminia/siti_archeo_villa.html.

allows, in fact, the valorization of buildings, monuments, historical-cultural sites, museums: an Augmented 3D vision device (such as Epson Moverio or Google Glasses devices) can be used to guide visitors, also implementing different languages, to provide textual and multimedia information units with explanatory or compendium aims; to provide preliminary, during, or after a visit, *ad hoc* materials, or multimedia products realized in progress by visitors themselves, through the presence of integrated tools (video, photography and movie registration). Through this technology, it is also possible to provide visitors with products that complement the existing and visible artifacts, such as the virtual recreation of parts and sections that no longer exist, or projections of future achievements as well as scenarios related to different eras and different contexts.

Moreover, the use of Augmented Reality devices can be experienced as a working tool for all those who, at different levels and with different specializations, must operate on components of the cultural heritage, designing and implementing recovery, usage plans, equipment, fruition paths. Such an application can be extended to all areas of art, with the creation of multimedia that complement and enrich the viewing experience, integrations for users with special needs, such as audio reading, magnification, integration with imaging, and the possibility to choose between different visualizations.

The ideas presented are applicable also for art pieces, both in terms of applications that belong to the recovery and restoration, classification/cataloguing, exhibition, promotion and "communication" of the work and/or the collection of works and their fruition. More specifically, for this type of artifacts, it is interesting to investigate the way in which the Augmented Reality can enrich the vision of the works of art itself, allowing not only to visualize related information, but also, for example, stages of restoration, layers and modifications, with respect to artistic styles, gallery of correlated works, historical-documentary contributions that relate to the artist, his biography and/or his/her execution techniques. These potentials have obvious repercussions both in the recognition and valorization of the artistic heritage, both during the use and sharing of it. Consider, as already mentioned, the possible integration with mobile devices through the use of last generation Q code.

Technology and Artistic Expression

Art does not consist in representing new things, but in representing them in a new way
("L'arte non consiste nel rappresentare cose nuove, bensì nel rappresentare con novità.",
Ugo Foscolo)

Art has always had characteristics of innovation and one of its most obvious functions is to dazzle viewers, creating new scenarios, placing the public in contact with new feelings as well as hidden and unconscious emotions. A fundamental aspect of the admixture of art and technology is the ability to explore new means of artistic expression. In fact, technology and algorithms underlying the A.I. models provide artists other means of expression, which have given rise to new types of art. Think of the artistic performances that make technology the focus to express the creative thought. For example, Virtual Reality technologies, closely related to A.I. for the management of systems, allows one to trick the mind, creating worlds and expressions that engage the mind of the user, retracing the sensations, the emotional states, and the artist's creative process. They range from applications such as video mapping, simultaneously available to more viewers, to more intimate applications individually livable. Even the robotics and the web 2.0 are widely used in new expressions of art. In 1950, there was a first artistic achievement whose focus was the technology, with the work of Ben Laposky and Manfred Frank, who created an "oscillogram" getting through a mathematical formula as a basis for graphic projections, creating special distortions. Laposky and Frank gave way to digital art, which also includes electronic music, or the so-called ASCII art (which uses images produced on the basis of seven-bit character encoding commonly used in computers). In the United States, in the sixties, Rauschenberg and Billy Kluver began the EAT (Experiment in Art and Technology) movement. From this period, the declinations of digital art have become increasingly numerous, too many to be all considered in this work. Just as an example, therefore, we will refer to some significant artists who have well played the communion of Art and Technology. We want to recall, for example, Stelarc, artist of the Posthuman Art, which is based on the idea of the obsolescence of the biological body, which can extend through the mechanics or electronics. In 1997, Stelarc comes to be implanted a third mechanical arm. As a proponent of extreme performance, based on interactive robotic and virtual reality systems, Stelarc is a key reference point for artists who aim to use technology as a means of expression.

There are, then, art trends such as the Transgenic Art: its great exponent, Kac, is remembered for Alba, a transgenic rabbit that becomes luminescent when illuminated at a particular electromagnetic frequency.

There are other artists, more or less "extreme", also Italian, we must remember. For example, Davide Coltro, a pioneer of the electronic picture, Nicola Evangelisti (Italian Light Art), Alessandro Brighetto (contamination between visual arts, physics and biology). And yet: musical and theatre performances; artists such as Mahnaz Esmaeli, Iran, who has applied the technology to the story of the artist; or even Marcel-li Antunez Roca, one of the founders of "La Fura dels Baus", one of the leading interpreters of the international cyborg scene. We cannot forget Bill Viola, one of the greatest exponents of Video Art with other artists such as Nam June Paik and Bruce Nauman, whose works have been exhibited in the major museums and art galleries, from MoMA to the Venice Biennale.

Art and technology give birth to such a large number of remarkable works that, at the turn of 2009 and 2010, the Victoria and Albert Museum in London involved the entire city, through the exhibition "Decode: Digital Design Sensations", that collected a great number of works of cutting-edge, emerging or otherwise artists, from all over the world.

The purpose of these forms of art is not only the creative expression in itself, but also between art and experiment, which stimulates the emergence of debates on the effects of technology on society.

Technology, Brains and Art: The Scientific Investigation

Among the latest technologies, we want to dwell here on those that have allowed significant advances, and promise interesting developments in the study of the interaction between art and the brain. In particular, we're refer to the recent Brain Imaging techniques that are not only important tools of research, but also suggest the possibility of using and developing tools that open up new futuristic artistic frontiers.

Recent advances in Brain Imaging triggered the opportunity to observe the brain in action (the so-called "living brain"), providing the opportunity to explore and assess real-time reactions (mental states, attention, involvement, stress) of users to specific stimuli. In fact, thanks to MEG, fMRI and EEG, researchers can study the response to specific stimuli in real time during an experiment performed in either virtual or real mode. Some of these technologies however, are very expensive and invasive (think to MRI—fMRI), or limit the movement (it is the case of the helmets used for medical EEG), or provide uncertain results, such as the Eye-tracking technique (about which discussions focus on the duration or the number of times that the subject pays attention to an object and on the value given to it). Studies on the cognitive aspects however, are largely based on Brain Imaging techniques (especially fMRI) and recently also on the analysis of electroencephalographic signals detected by medical or low-cost devices (the so-called Brain Computer Interface, BCI). Both the EEG and fMRI scans allow the observing of distributions as well as properties of brain activity at the topography level and respect the time, but due to its low cost, when compared to other methods, and thanks to the high temporal resolution, the EEG is a very promising approach in the evaluation of the reaction of individuals to stimuli, especially in a time perspective.

Born within the branch of Computer Science that studies models, methods and tools for interaction between man and machine (HCI, Human-Computer Interaction), the EEG-based BCI (Brain Computer Interface) offers the possibility to interact with a computer through brain signals and at the same time allow one to detect the response of individuals to specific stimuli in real time, thanks to algorithms and software for recording and interpretation of brain rhythms. Largely based on models of the A.I. initially designed for and used mainly in entertainment for the interaction within virtual environments and computer games, they are now also widely used by the scientific research community for applications and investigations in various areas. A BCI device (Allison et al. 2007) is a hardware/software system that collects electrical signals or other manifestations of cerebral activity and transforms them into digital data that a computer can understand, process and convert into actions and events. Such devices establish, in other words, a direct communication between the brain and an external device, such as a computer, based on the reading of electroencephalographic signals.

The success of BCIs is mainly due to the low cost, the high temporal resolution, the WiFi connection that allows individuals to feel relaxed, to reduce the anxiety, and to move freely in an experimental environment or play, acting and interacting as they would in the absence of the device, in a very natural way.

A BCI device records several brain frequencies grouped into rhythms: the alpha band (7–14 Hz), relating to a state of relaxation, meditation, contemplation; the beta band (14–30 Hz), associated with active thinking, with the attention to the solution of concrete problems; the delta band (3–7 Hz), frontally detected in adults, posteriorly in children with large waves and activities associated with sustained attention (Leeb et al. 2006); the theta band (4–7 Hz), generally attributed to emotional stress, such as frustration and disappointment; the gamma band (30–80 Hz), usually set in relation to the cognitive interpretation of multi-sensory signals.

BCI devices allow us to investigate the relationships between art and brain either from the point of view of the artist, during the process of creation of a work, and from the point of view of the public, while experiencing art. The interest is therefore to investigate the mechanisms of creativity and at the same time deepen the individual reactions during the observation of simple and complex visual stimuli. The brain functions of interest for this research (mainly responding to visual, auditory or cognitive stimuli) are related to the frontal cortex, because the visual system is the first perceptual system involved in cognitive processes. The area of the brain most involved is, in fact, part of the cortex called V1, designed to distribute the signals to different specialized areas of the brain.

If the assets, in color perception (Shapley and Hawken 2002) and in decision-making and memory functions (Barbas 2000; Vanrullen and Thorpe 2001), were already known, the various researches conducted allowed us to establish a link between the prefrontal cortex and the aesthetic experience, both in viewing works of art in their creation (Cela-Conde et al. 2004). In addition, studies of Hideaki Kawabata and Semir Zeki (Kawabata and Zeki 2004) revealed that the orbitofrontal cortex area is also involved in the judgment of the beauty of a painting. The same kind of brain activity was found even when the work is contextualized, for example through textual descriptions or multimedia. It is possible to therefore conclude, that the orbitofrontal cortex area performs the evaluation of sensory stimuli and that this analysis is significantly important in the evaluation of a work of art. Therefore, we can achieve not only to affirm intuitively, but also experimentally, that art is conceived by the artist and the object of observers' experience due to the neural activity of the brain, which create the aesthetic experience.

The purpose of our previous and ongoing research is to investigate tools and methods to collect, analyze and interpret the data recorded by the individuals during the observation or the creation of a work of art. To detect conscious and unconscious feedback, we used EEG-BCI devices, allowing freedom of movement. The focus of the research was placed on the cognitive and emotional response to basic and complex musical and visual stimuli, with the aim of transferring the results to reinforce the experience of art. The assessment of mental states and emotions was based on the valence/arousal model (Russell et al. 1989) that originates from the cognitive theory.

In the field of musical arts, to test the generalization ability of the selected EEG pattern and the associated emotive label, the emotional responses to physiological stimuli have been elicited using music and sound from the International Affective

Digitized Sounds database (IADS[6]). The results showed that mental activity measured in the front of the scalp distinguishes the valence of the musical emotions. In further experiments, participants were exposed to visual-perceptual, semantic or conceptual priming stimuli (Wiggs and Martin 1998), to assess the cognitive and emotional response in the context of museums of Visual Arts (Banzi and Folgieri 2012). Compared to the control groups (neutral stimulus, and the absence of stimulus), participants underwent stimulation and showed an increase in levels of attention at the questions related to the given stimulus, revealing a reinforcement of the memory mechanisms. These studies represent part of a wider interdisciplinary research project (Bait 2013, Calore et al. 2012, Folgieri et al. 2013) aiming to evaluate the response of humans to visual, auditory and perceptual stimuli, measured with classical methods used in Psychology and Cognitive Science and with innovative brain imaging methodologies, such as, in our case, the EEG and especially the BCI devices, both for the estimation of the reactions in a real environment (i.e. not an isolated laboratory, but in an environment also affected by external "noise"), both for the possibility of following development of applications for entertainment and education purposes, based on the research results.

The results of these studies have provided important insights about the devices and applications most appreciated by participants to the experiments, providing information used to design and implement some bridge solutions between the research and analysis of mental states induced by art, and the analysis and application of information technology to artistic creation. This is the case of the study on the reproduction of conscious music melodies using EEG signals provided by individuals (Folgieri and Zichella 2012), allowing the creation of a tool that enables users to create sounds and melodies with the only support of their brain rhythms.

A further application designed to allow both the exploration of new artistic means and the brain mechanisms involved in the artists' creative process, consists of a workbench named BrainArt, on which the present study is mainly focused.

3   The BrainArt Workbench

The BrainArt workbench was born especially for entertainment purposes, as an application that provides artistes (and non-artistes) of an area the ability to define shapes, symbols and colors to be used in the subsequent creation of a visual work, generated directly by their brain rhythms. The artiste, once equipped with the BCI device, is free to express her/his moods and creative insight that is interpreted from the computer thanks to A.I. algorithms (a neural network) classifying and interpreting brain signals in the light of the meanings attributed to different rhythms in the literature and supported by the previous studies described (arousal, excitement, anxiety, attention, relaxation, etc.). In this way, the artist is allowed to "paint" with

---

[6] http://csea.phhp.ufl.edu/media/iadsmessage.html.

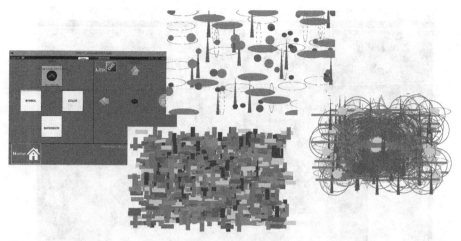

Fig. 4.1 The application and some examples of the created works

the thought, after an initial setting phase to adapt the software to personal parameters (Fig. 4.1).

In addition, to allow users to test their creative ability, the application enables the recording of the brain rhythms detected during the creation of the work, so that it is possible to analyze the patterns (*a posteriori*) to investigate the underlying cognitive processes.

3.1   *Technical Specification*

The development was conducted using Java, to allow the portability across different operating systems and to benefit from object-oriented programming.

The architecture has been designed to allow a pleasant and intuitive user experience (the quality of which has been submitted to the judgment of a group of heterogeneous users by age, sex, education, and artistic inclination) and the workbench is equipped with an easy-to-access online help. Furthermore, the graphical interface and the application itself are both designed to be used both in traditional and tablet PC, as well as touch devices in general. The implementation effort is concentrated on maintaining a low computational level, to allow the use of BrainArt also on limited capacity devices.

For the brainwave detection, the application interfaces to Neurosky MindWaveTM, chosen for its affordability, good technical support, the availability of free SDK provided by the manufacturer, and mainly because it is created for entertainment purposes.

The MindWave (an evolution of the MindSet) is a device developed by Neurosky, available to the public since 2011. It consists of a dry sensor mounted on a semi-rigid support. The device is composed of a passive sensor positioned in Fp1

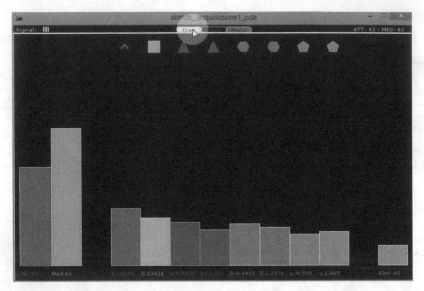

Fig. 4.2 The graph tab, showing, in a *bar* graph, the intensity of cerebral rhythms

(left frontopolar) and from a reference sensor, positioned on the earlobe, used to subtract the common ambient noise through a process known as common mode rejection. The Neurosky Mindwave registers brain activity frequency between 3 and 100 Hz, automatically filtering out the rhythms in high/low alpha, high/low beta, mid/low gamma, delta and theta. In addition to these rhythms, the device also captures the eye blink and, with the usage of proprietary algorithms, the level of attention and relaxation (e-sense™ meters) expressed as a percentage. The analog to digital converter (ADC) present in the MindWave is composed of a sampler to 512 Hz and from a 12-bit quantizer. The BCI device uses a wireless connection to the computer. The data is collected and recorded by a specific protocol that establishes a connection following a client-server model, communicating via JSON type string.

3.2 The User Interface

The workbench is organized following a "tabbed browsing" model. Each window will start only when it is invoked by the user. The tabs are three: Drawing, Graph, Menu.

The Drawing tab allows one to visualize the work in progress.

The Graph tab displays a bar graph presenting the intensity of the individual waves detected (Fig. 4.2).

The Menu tab allows one to change the settings for the symbols, colors and size of the objects that the artiste wants to use to create his drawing. The user can also

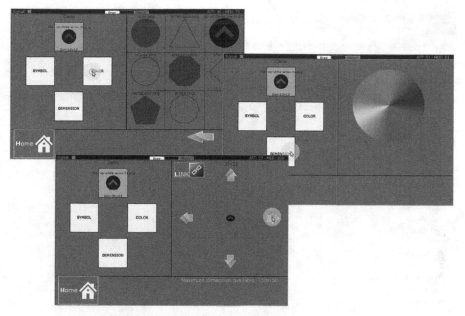

Fig. 4.3 Screenshots of the personalization possibility given in the Menu tab

define symbols, shapes and colors as he/she wishes, according to his/her own taste and creative inspiration (Fig. 4.3).

The Design tab represents an ideal "canvas". Through the Play/Pause button, the user can stop drawing to resume work later. The Save button allows one to save the work at any time desired. However, at intervals determined by the users at the beginning of the session, intermediate images (and the related EEG data) are also saved. The saved image is composed by the drawing and a header with date, time, and user name.

Each symbol/object, chosen by the user during the customization phase, is associated with a brain rhythm. Once the individual choices are made, at the program startup, the Design tab begins to show the creation in progress. The frequency with which a symbol is repeated is determined by the intensity of brain activity, related to the rhythm linked to the symbol. The user is, thus, able to choose, in a relatively conscious way, the proportion with which the symbols appear in the scene.

Through an additional software (BrainArt-Train) related to the main one, the artiste, wearing the MindWave, can work out in advance to be able to impose a higher amplitude for certain rhythms.

The BrainArt workbench is multiplatform and multilingual and allows, as already mentioned, the registration of brain rhythms, blink events and the e-sense meter values collected, in the formats GDF, EDF and CSV. The saved file not only allows to play (re-interpreting the data) all the intermediate images produced by a user in the process of creation of a work, but it also allows to analyze the collected data for research purposes (Fig. 4.4).

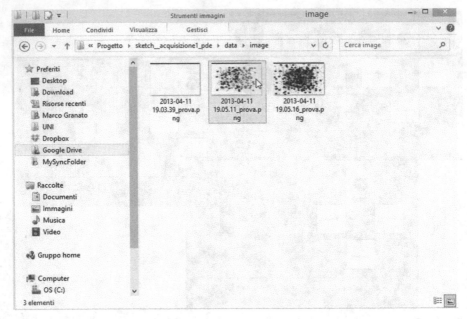

Fig. 4.4 Screenshot of the saved pictures

3.3 The Experimental Study

Given the possibility of using the software to register the individuals' rhythms, we decided to set up an experiment, keeping in mind the "simple" aim of analyzing correlates and antecedents of the creative act.

We wish to underline that our objective is not to study the neuropsychology of the creative act through Brain Imaging equipment, but to keep the occasion to observe the individuals' creativity attitude when free to express it in a game environment. The study concerns, so, the unconscious mechanisms underlying the moment in which the unconscious creative act meets the conscious evaluation of the resulting product, that is, when the user observes her/his creations and chooses the one she/he judges more representative of her/his insight. In other words, we wish to compare the act of creation (intimate and spontaneous by nature) with the neuroaesthetic judgment (influenced, instead, also by external factors, such as one's own culture, general aesthetic, and so on).

Experimental Design

The experiment consisted in the following procedure, expressing what we wrote above:

1. participants were informed of the objective of the study, that is, to register their brain rhythms by a BCI device, during the creation of a drawing expressing their creative idea;
2. participants were instructed on the following experiments and on the BrainArt workbench features;
3. the participants did the experiments;
4. the participants were asked to answer a post-test (TAS) consisting of a cognitive questionnaire to assess their creativity.

Each experimental session had been forerun by an individual setup phase, during which every individual was asked to choose (or create) symbols, colors, shapes and min/max dimensions to associate at each cerebral rhythm, to prepare the workbench to provide the means (such as preparing palette, paint brushes and sketch in traditional painting) to express the personal creativity. Once done, the participants wore the BCI and started the BrainArt workbench creating the artwork.

As told, BrainArt registers EEG data and intermediate images created during the session, as well as a movie of the overall creation. To make the registration uniform (as much as possible), in our experiment, the intermediate images have been saved with an interval of one second, for all the participants. Obviously, the duration of the experiment varies for each individual, because the creative expression depends on personal insight and behavior.

To obtain a spontaneous creative act, none of the participants used the training software. All the individuals, in fact, directly used the BrainArt workbench.

At the end of each experiment, that is, when a user decided to stop the creative session, every participant was asked to choose, among all the saved images, the one that represented his/her creative idea the most. So, only in a second moment, will the individual express his/her own aesthetic judgment, choosing the image that is closer to his/her creative idea (in this phase the conscious and unconscious processes meet).

After, the TAS questionnaire will been submitted to each participant. The Tellegen Absorption Scale (Tellegen and Atkinson 1974) is a 34-item multi-dimensional measure that assesses the imaginative involvement and the tendency to become mentally absorbed in everyday activities. Absorption represent a sort of attentional deployment, now interpreted and codified as a personality trait, and has found its way into research areas as diverse as aesthetics, hypnosis, psychopathology, religious behavior, altered states of consciousness, intelligence and dissociative states. Individuals higher in trait Absorption appear to manifest a fundamental need to experience the natural world in a way which totally engages attentional resources in an intrinsically self-rewarding way. Tellegen and Atkinson describe the absorptive experience as an "'allocentric' perceptual mode...involving 'totality of interest', and openness to the object in all its aspects with all one's senses" (Tellegen and Atkinson 1974). Furthermore, this attentional style is total, "involving a full commitment of available perceptual, motoric, imaginative and ideational resources to a unified representation of the attentional object" (Tellegen and Atkinson 1974). We argue that abso"ption could be an important factor in guiding the balancing or rebalancing of conscious and unconscious processes that the use of BrainArt implies. Indeed, people high in absorption should show a higher ability to totally converge their attention to the task, thus letting the brain free to express ideas and concepts preventing distractions by external and internal stimuli.

The experiments were conducted among voluntary individuals. The population consisted of 10 males and 10 females, of different age (age range 16–60, equally distributed) and education level. A quarter of the participants, equally distributed by age and level of education, received an education in art (Art and Humanities studies).

During the experimental session, the participants sat at a distance of 50 cm from a screen, wearing the Mindwave device. They were asked to sit in a comfortable position, to concentrate on the creative idea (they were observing the work art composing, on the screen, during the experiment) and avoid talking or moving during the creation of the artwork if possible, in order to avoid the influence of Electromyography (EMG) signals in the collected data.

Results

The experiment provided some interesting indications concerning the insight moment and its correspondence with the aesthetic choice. The observed data and the following analysis indicated some remarkable correlations between the creative act and the cerebral rhythms involved, confirming also previous works in literature (Doppelmayr and Weber 2011; Kwiatkowski 2002; Razumnikova 2007; Srinivasan 2007). Moreover, a difference between individuals who received or did not receive an education in art also emerged.

The analysis of the collected data was divided into three stages:

1. the Behavioral data analysis, based on the findings deriving from the questionnaire submitted to participants,
2. the EEG signal analysis conducted to individuate some patterns revealing that particular rhythms are involved in the creative processes,
3. the EEG data statistical analysis conducted to confirm (or confute) the previous findings.

Before analyzing the EEG signal, here is already possible to highlight the really interesting results of the post-experiment survey, described in the next paragraph.

Behavioral Data Analysis

The absorption level, as measured by the TAS questionnaire, showed interesting correlations with the concentration score and the EEG dynamics. In particular, it a direct correlation between the TAS score and the concentration level achieved at the moment of the creative act as measured by the BCI software (Parson is $=0.387$; $p=0.003$) appears clearly. Interestingly, the highest absorption score (TAS score

=24) was reported from the subject with the highest concentration level (64%). This data appeared to be independent of the subjects' artistic education. Indeed, though the ability to achieve a high level of concentration appear to be independent of artistic education, the TAS score seems to be able to predict the ability to achieve high focused attention state of mind. Indeed, artistic education probably guide and modulate conscious process, but it doesn't seem to influence the automatic and implicit processes required to modulate attention mechanisms via BCI and neuro-feedback systems.

Furthermore, the EEG theta band dynamic appears to be differently modulated in subjects with different level of absorptions. In particular, subjects with high level of absorption showed broader temporal theta synchronization windows, whereas low absorption subjects appear to focalize attention just near the creation act. Indeed, they show a theta peak just few second before "creation", while high absorption subjects increases (unconsciously) theta power some minutes before. It is then possible to state at least two main suggestions:

- Theta synchronization is probably the main EEG correlate of creativity.
- Subjects high in absorption trait are able to get into a prolonged pro-creativity state of mind, maintaining for several minutes a high level of focused attention (theta synchronization), and then showing a higher concentration peak just before the creative act.

EEG Signal Analysis

The data collected has been analyzed using different software tools. Particularly, we used the WEKA (Hall 2009) software and *ad hoc* implemented MATLAB™ code. We recall that the signal transmitted by the device[7] is already filtered to remove the 50 Hz frequency band related to electric power equipment. Moreover, it detects eye blinks, thus avoiding a specific filtering work. Other motion related activities should be detected with ad hoc signal analysis, that it has not been considered, since from a visual inspection the signal in general did not show these typical features.

For every participant, we conducted the analysis of the different brain rhythms to detect the maximum brain activity and verify if it occurs correspondingly at the registration time of the picture chosen by each user as the most representative of

---

[7] The MindSet provides information about the wearer's eSense™ Attention and Meditation levels, along with the wearer's relative brainwave/EEG frequency band powers, such as delta, theta, alpha, beta, and gamma waves. It also provides information about the quality of the signal/connection. Lastly, it can provide the measured raw wave (EEG) samples.

The data values collected by the Neurosky Mindwave represent the current magnitude of 8 commonly-recognized types of EEG (brain-waves). This Data Value is output as a series of eight 3-byte unsigned integers in little-endian format. The values of the eight EEG powers have no units and therefore are only meaningful compared to each other and to themselves, to consider relative quantity and temporal fluctuations. Otherwise, the Neurosky provides a formula to convert data into voltage. The formula is: [rawValue * (1.8/4096)]/2000.
This is due to a 2,000x gain, 4096 value range, and 1.8 V input voltage.

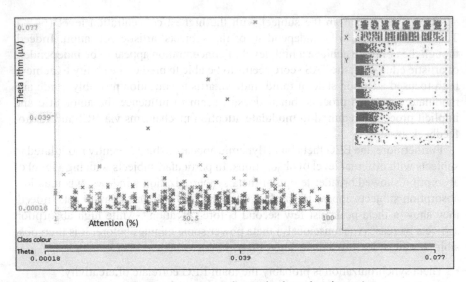

Fig. 4.5 Theta and Attention data from a user who received an education to Art

her/his creative idea. From a first visual inspection, we concentrate our analysis on the theta, high beta and mid gamma rhythm and on the attention level.

We also performed FFT for each second of data and computed the correlation coefficient, to verify if couples of bands are related.

In general, in all the participants, the creative moment (defined as the one declared by the person who has chosen, among those registered, the image that best represents her/his creative idea) is accompanied by the maximum intensity of the theta rhythm. In fact, the theta rhythm increases in intensity in conjunction with what has been indicated by users as the "creative moment". Also, the activity of theta intensifies, and in some cases, in the instants preceding or following the creative moment, it is possible to note an increase in intensity and activity rhythms as well as low beta and mid gamma rhythms. This latter phenomenon occurs only in users who had received education to art.

In the signal analysis, the value of attention level (in percentage) provided by NeuroSky has been also considered to evaluate if, and how, it is related to the above rhythms. For the majority of participants in the experiment (80%), the level of attention detected when the theta rhythm has the maximum intensity is beyond the 40%, so it is high enough (a subject has even the value of 74) to be considered. The 20% of individuals presents, however, significant value, since they are around the 30%. The link between the level of attention and theta rhythm appears to be independent of whether or not individuals received art education.

As an example, since it is not possible (for space limitations) to show data from all the participants, in the following figures we show results from a participant who received art education (Fig. 4.5).

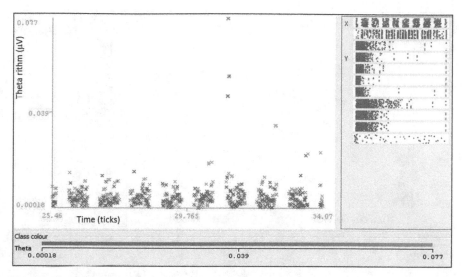

Fig. 4.6 Theta rhythm from a user who received an education to art

For 50% of the participants the theta rhythm shows the maximum intensity in the narrow time around the one registered for the image chosen as representative of their creative act. The interval, in this case, ranged from a few seconds up to 1 min before and a few seconds up to   min after the time corresponding to the selected image.

For the remaining 50% of the participants, the interval ranged from around 3 to 6 min before and after the chosen image, occurring very closely, however, to the time indicated as "creative" by the subjects.

The intensity of theta rhythms appeared independent from artistic education received or not by individuals (Fig. 4.6).

For the 80% of users (not provided by artistic training), the beta rhythm (particularly the low beta) did not seem related to the theta rhythm. Only for the 20% of users did (all provided by artistic education) it show its maximum intensity from a few seconds to 2 min before or after the instant of maximum intensity detected by the theta rhythm. In the latter case, the incidence of cases in which the intensity of low beta is maximum before that of theta prevails, almost to denote a design activity before the more pure "creative abandonment" (Fig. 4.7).

For the 70% of participants, the mid gamma rhythm did not reveal relationships with the theta rhythm. For the remaining subjects, the intensity of the mid gamma rhythm showed its maximum before the maximum intensity of theta (from one up to four minutes before) as to indicate a memory recall of a planning activity guiding the creative moment. Also in this case, this fact occurs in participants who received artistic education (Figs. 4.8 and 4.9).

As said before, for each individual, the FFT was also performed for each second of data and the correlation coefficients were computed for the couple of rhythms "theta-low beta", "theta-mid gamma" and for "theta-attention". The values of the

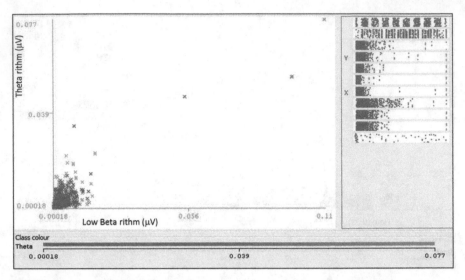

Fig. 4.7 Theta and Low Beta rhythms from a user who received an education to Art

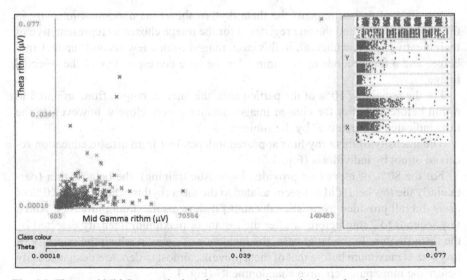

Fig. 4.8 Theta and Mid Gamma rhythms from a user who received an education to art

correlation coefficient results were significant only for the couple "attention-theta" (ranging, for all the participants, from 0.637 to 0.856), while for theta-low beta and theta-mid gamma, only for the individuals who received an artistic education, confirming the previous findings. In Table 4.1 below, we summarized the results, grouped for couples of rhythms and correlation coefficient ranges.

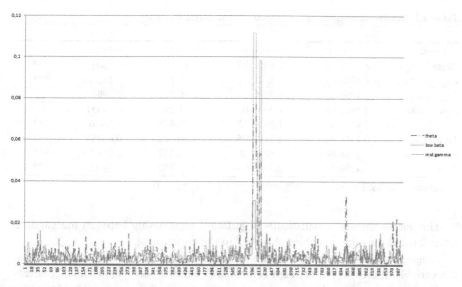

Fig. 4.9 Theta, Low Beta and Mid Gamma rhythms from a user who received an education to art

Table 4.1 Correlation coefficients (for all the participants) for the considered couples of rhythms and grouped for ranges

Rhythms	Correlation coefficient range	Artistic education
Theta-Attention	0.637–0.856	Non influential
Theta-High Beta	0.189–0.253	No
Theta-High Beta	0.465–0.523	Yes
Theta-Mid Gamma	0.148 0.233	No
Theta-High Beta	0.473–0.583	Yes

EEG Data Statistical Analysis

The first statistical analysis performed on EEG data consisted in a regression data analysis.

We decided to analyze data from each individual before singularly, and after as a whole, to assess which variables have the most influence on the process considered, and in particular, what the links between the rhythms and the level of attention are. For the analysis of the data, we used the statistical software R[8]. For each subject, a regression analysis was carried out, using the linear model function (lm) for the variable *attention*, an eventual restricted model and the calculation of all the correlations between the variables.

The variables used are: delta, high alpha, low alpha, meditation, low range, low beta, theta and mid gamma.

---

[8] http://www.r-project.org/.

Table 4.2 Results from regression analysis on all the collected data

	Std.	Error	t value	Pr(>|t|)	
Estimate	5.067e+01	9.565e-01	52.978	<2e-16	***
Delta	−8.885e-06	7.032e-07	−12.635	<2e-16	***
High_alpha	−5.438e-05	8.861e-06	−6.137	9.11e-10	***
High_beta	−1.466e-05	1.739e-05	−0.843	0.39947	
Low_alpha	−2.194e-05	1.694e-05	−1.295	0.19531	
Low_beta	2.411e-04	1.940e-05	12.428	<2e-16	***
Low_gamma	6.352e-06	1.240e-05	0.512	0.60857	
Mid_gamma	1.349e-04	2.628e-05	5.132	2.99e-07	***
Theta	−3.423e-05	2.564e-06	−13.353	<2e-16	***

Significance codes: 1 = .01; . = 0.05; * = 0.01; ** = 0.001; *** = 0

The output showed a statistically significant relationship between the variable attention and the other variables observed in the experiment.

Judging and evaluating the overall total experiment it can be seen, excluding the variable intercept (the intercept is included in this model too, by default, unless explicitly suppressed with ~− 1 in the regression formula) that there are at least two variables that are significant or very significant in almost all the data collected from participants, namely low beta and theta.

In fact, low beta and theta are the only two independent variables that show the highest (***) significance.

We then decided to add the asterisks of significance provided as the output of the regression function, assigning a value of 1 to each asterisk and value of 1/2 to every ".", to consider the variables showing a greater significance. We therefore proceeded with the regression on the unified data.

We decided to evaluate all the data in their entirety, that is, by combining all the twenty series of data collected, to verify if the behavior of the variables, as the number of data (even if from different individuals), were the same as recorded for individual participants.

The formula used for the regression is the same used to analyze the twenty subjects in a separate manner.

The values obtained are shown in the following Table 4.2.

Comparing this output with the previous ones, we obtained a confirmation of the findings and significance previously found; non-significant or less significant variables are, even in this case, represented by high beta, low alpha and low gamma. We note, finally, that the variable mid gamma is very significant. This may indicate that the single data (from one individual) cannot provide relevant information for our research but, when we aggregate all the data, mid gamma is a variable that explains the cognitive dynamics of the subject and the involvement of attentive mechanisms in the creative process.

Discussion

The analysis of the data collected during the experimental phase was conducted using three different approaches to evaluate the common conclusions. We conduced, in fact, a signal analysis and a statistical analysis of data and, in addition, a cognitive analysis of the data collected from the test (TAS) submitted to the participants.

From the analysis of the intensity of brain rhythms, the rhythms more involved, i.e. those that showed significant differences in the neighborhood of time corresponding to the image chosen by the participants, are three: the theta, the low beta and the mid gamma rhythm. Both from the point of view of the analysis of the signal, and from the point of view of statistical significance, the theta rhythm showed an increase in intensity and activity in all the individuals observed. The other two rhythms are apparently relevant only in the case of users who have received an artistic training.

We also considered the value of attention level provided by Neurosky, to evaluate how this was connected to the rhythms mentioned above. The results show that, in the creative process, attention plays an important role, especially in subjects who received an art education.

Having invited the participant to choose the image (among the recorded ones) *a posteriori* in which they recognize their own creative act, the choice was undoubtedly influenced by personal aesthetic taste, so it is not possible to detect a precise moment per second (even less per millisecond) corresponding to the creative act. Moreover, since we wish to analyze a macro-phenomenon, we must keep in mind the fact that the choice made in retrospect may also differ in minutes compared to a peak of creativity (intensity of theta brainwaves) recorded with the EEG. This time interval, remarkable when compared to the brain response to stimuli (of the order of milliseconds) is acceptable, however, for the type of phenomenon considered and, especially, considering how the experiment has been set. In fact, we want to recall that to search for the emergence of patterns eventually connected to a creative activity, the participants were asked to choose which was the image that best corresponded to the work they wished to realize and, therefore, at this stage we cannot have a human response digitally precise, and in perfect synchronization, with the results of the data collected.

However, the investigation has significant value because the data indicated patterns that, in the future, would allow us to reverse the experiment, that is, selecting the "most creative" image on the basis of the patterns detected by brain rhythms and then submitting it to every participant, asking him/her to express, in a scale of fixed values, how much the image was close to his/her original creative idea.

4   Conclusions and Further Developments

In this work, we presented our point of view on the relationship among Art, Brain and Technology. Particularly, we kept the occasion to present the BrainArt workbench developed mainly for entertainment (and, why not, educational) purposes, but also gave us the opportunity to perform a high-level analysis of the creative process by a Cognitive Science perspective.

In cognitive science, the functioning of the brain and the achievements of art are considered together to explain our aesthetic experience. The case of color processing may be considered particularly relevant. For instance, it is possible to explicate why sometimes the reading of some images may depend on a subjective interpretation, based on the personal and sentimental response despite to perceptive cues, as colors and their contrasts. Indeed, colors take on different meanings in terms of aesthetics, and it is not unlikely that they have the effect of evoking feelings, emotions and moods.

For example, it is a fact that black can hardly put ourselves in a good mood, while yellow can elicit an impression of youthful freshness and vitality, and each of us has specific preferences about. In red, the color of blood and fire, it is implicit in fact, the sense of heat, of vital energy, but also of danger. The blue instead of red symbolizes the opposite and it is cold, detached and rational. The green dominant color of the vegetable world, meadows, nature, gives us a sense of calm and confident expectation, as in the case of the traffic light. Further than individual colors, importance must also be given to the combination of different colors, that is why some colors, next to each other, give a sense of oneness and harmony, while others do not.

Here there are differences as great as that in the judgment on the isolated colors: there must be a sort of color harmony. It's hard to believe that our brain uses a scientific criterion, quantitatively defined in the same way as a colorimeter does to determine the exact nature of a color. Moreover, any combination of colors made by nature generally seems harmonious and acceptable. In nature, however, the endless game of shadows, light and shade, and color gradations, greatly reduces the possibility of sudden contrasts and juxtapositions. It is likely that when facing nature or human products, our mind works differently, since in the latter case we tend to judge more or less successfully a combination of colors, due to a supposed higher degree of freedom. Subsequently, color patterns found in nature seem to be processed in an unconscious way, why human-made patterns stimulate a consciousness-driven process. Though both unconscious and conscious processes may modulate arousal and elicit emotions or feelings, a consciousness-driven evaluation may introduce idiosyncratic aesthetic and hedonistic effects. According to the French chemist Michel Eugene Chevreul, there is harmony along two main lines: when the colors are similar to each other, or, at the other extreme, when they appear to be in contrast. Regarding the harmony for affinity, in the case of adjacent colors, the centre color is supported and enhanced by its neighbors; shades of this type are also found in nature, as can be the change from red to blue in a sunset through all the shades in between. Another type of affinity is the one associated with the dominant color, also deriving from the habitual daily experience. Regarding instead the harmony of contrast, the most conspicuous example is the harmony of complementary colors.

When we perceive the natural world, this sort of intrinsic harmony unconsciously arise in our mind, and we don't need conscious considerations to appreciate it. Otherwise, observing man-made products, such as a painting, the balance between conscious and unconscious processes are much more variable and the related experience may change from time to time, and person by person. If this true when observing art, it should also true when producing art or, more simply, when we use our creative mind, or when canonical thought is replaced by divergent thinking.

Indeed, the creative act or insight is borderline phenomena, in which conscious and unconscious are mixed, finding a rebalancing, where attention is implicitly absorbed by a target that seems to fluctuate between a conscious and an unconscious representation, before emerging as a new idea or insight. In this perspective, that considers creativity processes as the consequence of the brain system to continually find a dynamic balance in order to achieve given aims (and this balance gives rise to the mind state that we experience moment by moment) the presented "simple" experiment, even if explorative and absolutely preliminary, can be considered paradigmatic. On the basis of the discussion of the results, can we state that it exists a theta, a beta/theta or a gamma/theta threshold able to define the transition from the impression to the expression? It seems that the beta/theta and gamma/theta relationship results significantly only when the individual has received an artistic education before. May this mean that when the subject is more aware of his/her own creative processes (having developed specific skills or knowledge), then the beta and gamma rhythms report an imbalance or "rebalancing" toward consciousness, that is, toward conscious processes? Future research works could consist in answer to these questions. At the moment, we can only state that deepening our understanding of the relationships between EEG dynamics and specific task performing is required, in order to really appreciate the functioning of the cognitive balancing process that give rise to dedicated state of mind.

Moreover, the application of the Cognitive Science approach to Art, entertainment and educational fields also represents a promising approach. In fact, future learning and entertainment environments will be cognitive-based, in order both to enhance the aesthetic and hedonic experience and to tune the environments to the user needs, preferences and skills. In this sense, simple BCI-based application, as BrainArt, will allow developers to implement neurofeedback systems aimed at enhancing in real-time the human-machine communication. It is also plausible that in a near future art developers will have the possibility to use similar systems in order to bypass consciousness censorship that might hinder creativity and related productions.

Acknowledgments The authors thank Miriam Bait for her precious help and comments in writing this paper.

References

Allison, B. Z., Wolpaw, E. W., & Wolpaw, J. R. (2007). Brain–computer interface systems: progress and prospects. *Expert review of medical devices*, 4(4), 463–474.

Andreasen N.C., (2005), The Creating Brain: The Neuroscience of Genius. *New York and Washington DC*: Dana Press.

Bait, M., Banzi, A., Folgieri, R., & Minetti, S. Intrducing a virtual reality eeg-bci and priming-based tool to make art interactive: a technological and linguistic challenge. *EVA 2013 Florence*, 152.

Banzi, A., & Folgieri, R. (2012). Preliminary Results on Priming Based Tools to Enhance Learning in Museums of Fine Arts. In *Proceedings of EVA* (pp. 142–147).

Barbas, H. (2000). Connections underlying the synthesis of cognition, memory, and emotion in primate prefrontal cortices. *Brain research bulletin, 52*(5), 319–330.

Baumgarten, A. G. (1735). Reflections on poetry. Berkeley, University of California Press.

Berenson, B. (1948). A esthetics and History in the Visual Arts.

Brandt, P. A. (2006). Form and meaning in art. The artful mind: Cognitive science and the riddle of human creativity, 171–88.

Breidbach, O. (2003). The Beauties and the Beautiful—Some Considerations from the Perspective of Neuronal Aesthetics. In *Evolutionary Aesthetics* (pp. 39–68). Springer Berlin Heidelberg.

Calore, E., Folgieri, R., Gadia, D., & Marini, D. (2012, February). Analysis of brain activity and response during monoscopic and stereoscopic visualization. In*IS&T/SPIE Electronic Imaging* (pp. 82880M–82880M). International Society for Optics and Photonics.

Cela-Conde, C. J., Marty, G., Maestú, F., Ortiz, T., Munar, E., Fernández, A., ... & Quesney, F. (2004). Activation of the prefrontal cortex in the human visual aesthetic perception. *Proceedings of the National Academy of Sciences of the United States of America, 101*(16), 6321–6325.

Changeux, J. P. (1994). Art and neuroscience. Leonardo, 189–201.

Csíkszentmihályi, M. (1990), Flow: The Psychology of Optimal Experience, New York: Harper and Row, ISBN 0-06-092043-2.

Doppelmayr, M., & Weber, E. (2011). Effects of SMR and theta/beta neurofeedback on reaction times, spatial abilities, and creativity. *Journal of Neurotherapy, 15*(2), 115–129.

Folgieri R. (2011). VR for cultural heritage valorization: a communication problem. In: *Proceedings of Electronic Imaging & The Visual Arts*. Firenze, Italy, 2011, p. 146–151, Cappellini, ISBN: 88-371-1837-6.

Folgieri, R., & Zichella, M. (2012). A BCI-based application in music: Conscious playing of single notes by brainwaves. *Computers in Entertainment (CIE), 10*(3), 1.

Folgieri, R., Lucchiari, C., & Marini, D. (2013, February). Analysis of brain activity and response to colour stimuli during learning tasks: an EEG study. In*IS&T/SPIE Electronic Imaging* (pp. 86520I–86520I). International Society for Optics and Photonics.

Freedberg, D., & Gallese, V. (2007). Motion, emotion and empathy in esthetic experience. *Trends in cognitive sciences*, 11(5), 197–203.

Gallese, V., & Lakoff, G. (2005). The brain's concepts: The role of the sensory-motor system in conceptual knowledge. *Cognitive neuropsychology*, 22(3–4), 455–479.

Geiger, M. (1928). Zugänge zur Ästhetik. Der Neue Geist Verlag.

Gombrich, E. H. (1959). The Concept of Style in *the History of Art: Lecture Notes [for] Fine Arts 190*, Spring Term, 1959. Harvard University, Department of fine arts.

Hall, M., Frank, E., Holmes, G., Pfahringer, B., Reutemann, P., & Witten, I. H. (2009). The WEKA Data Mining Software: An Update; *SIGKDD Explorations*, Volume 11, Issue 1.

Kawabata, H., & Zeki, S. (2004). Neural correlates of beauty. *Journal of neurophysiology, 91*(4), 1699–1705.

Kwiatkowski, J. (2002). Individual differences in the neurophysiology of creativity.

Leeb, R., Keinrath, C., Friedman, D., Guger, C., Scherer, R., Neuper, C., ... & Pfurtscheller, G. (2006). Walking by thinking: The brainwaves are crucial, not the muscles!. *Presence: Teleoperators and Virtual Environments, 15*(5), 500–514.

Norman, D. (1981). Perspectives on Cognitive Science.

Oliverio A. (2009) "La vita nascosta del cervello" Giunti Editore Firenze.

Pinotti, A. (2005). Introduction in Art in the Age of Visual Culture and the Image. Leitmotiv, 5, 7–10.

Razumnikova, O. M. (2007). Creativity related cortex activity in the remote associates task. *Brain research bulletin, 73*(1), 96–102.

Rizzolatti, G., Fogassi, L., & Gallese, V. (1997). Parietal cortex: from sight to action. *Current opinion in neurobiology*, 7(4), 562–567.

Russell, J. A., Lewicka, M., & Niit, T. (1989). A cross-cultural study of a circumplex model of affect. *Journal of personality and social psychology, 57*(5), 848.

Shapley, R., & Hawken, M. (2002). Neural mechanisms for color perception in the primary visual cortex. *Current Opinion in Neurobiology*, *12*(4), 426–432.

Srinivasan, N. (2007). Cognitive neuroscience of creativity: EEG based approaches. *Methods*, *42*(1), 109–116.

Stafford, B. M. (2005). Art and Consciousness: Methodologies. Innovation and Visualization: Trajectories, Strategies, and Myths, 1, 37.

Stafford, B. M. (2007). Echo objects: The cognitive work of images. University of Chicago Press.

Tellegen A. and Atkinson G., Openness to Absorbing and Self-Altering Experiences ("Absorption"), a Trait Related to Hypnotic Susceptibility, *Journal of Abnormal Psychology*, 83, pp. 268–277, 1974.

Vanrullen, R., & Thorpe, S. J. (2001). The time course of visual processing: from early perception to decision-making. *Journal of cognitive neuroscience*, *13*(4), 454–461.

Wiggs, C. L., & Martin, A. (1998). Properties and mechanisms of perceptual priming. *Current opinion in neurobiology*, *8*(2), 227–233.

Worringer, W. (1907). Abstraktion und Einfuhlung: Ein Beitrag zur Stilpsychologie. Universitat Bern.

Zeki, S. (1999). Art and the brain. *Journal of Consciousness Studies*, 6(6–7), 6–7.

Zeki, S. (2004). The neurology of ambiguity. *Consciousness and cognition*, 13(1), 173–196.

Chapter 5
Video Ergo Sum: An Artist's Thoughts On Inventing With Computer Technology In The Creation Of Artworks

Nathan Cohen

1 Introduction

The computer, while not a new concept, has in its modern form transformed the way we disseminate ideas, interact with one another and enhanced our capacity to acquire information. From the artist's perspective digital imaging presents opportunities for visual invention and challenges in how visual form is mediated. In my artwork I use the computer as a means to create imagery that would not be possible without its use, and that enables exploration of an artificially created space that enhances spatial awareness and challenges our perception of what we encounter. The computer enables the use of real time and recorded moving and still images to be embedded within artwork previously limited to still imagery and makes possible the fragmentation and reconstruction of the picture plane into multiple moving images with a remarkably high degree of resolution.

So why would being concerned with the way an image is presented be significant and how might considering this question help us to advance pictorial invention in a digital age? How we encounter the world is essentially a very personal act and we do this in a way that is both knowing (based on prior experience) and questioning (open to new experience). In choosing to make an artwork I am seeking to explore both of these perspectives and in doing so one of the big challenges is how to make an illusion of space appear real and to find a way to make spatially comprehendible what is in essence an invention. To achieve this it is possible to be creative with spatial arrangement in an artwork that, while clearly defined, is also open to interpretation resulting in illusions of space that are intriguing for the viewer, enhance engagement and challenge spatial perception.

There is a different sensibility to image generation on a computer compared with the articulation of visual ideas made by hand and a graphic medium. Our impulse to make marks is evident in the long history of image making dating back to the

N. Cohen (✉)
University of the Arts London, London, UK
e-mail: n.cohen2@btinternet.com

N. Lee (ed.), *Digital Da Vinci,* DOI 10.1007/978-1-4939-0965-0_5,
© Springer Science+Business Media New York 2014

earliest pictographs and petroglyphs to be found drawn and carved onto natural surfaces using little more than pigment and rock. Programs for drawing digitally onto a screen or tablet have been around for a while but the computer screen offers a singular format for creating and presenting these images. While our modern mathematically proportioned paper formats do in some part offer a parallel, the range of format and adaptability available to the artist using different media to create their work is substantial when compared to the commercially available screen in its present form, particularly when wishing to create images not contained within a standardised shaped surface.

When Brunelleschi stepped back into the door whose frame outlined the view of the building he wished to represent, the fixing of the edge of the observable space defined it as a picture plane onto the three-dimensional space beyond, marking a moment where the framing element established a means to create a spatially measurable illusion of reality (Kemp 1990). However, it also defined a boundary between the real world of the viewer and that of the image and ensures our perception of what is depicted remains consciously an illusion.

We take this invention for granted today and are familiar with looking at artworks that represent to us a likeness of the world we are familiar with in a way that is spatially convincing, allowing us to engage with the illusion of space as an extension of our real world reality.

The creation of an illusion of space that can be interacted with on a human scale can be seen where the properties of art, architecture and effects of light combine. Historical examples of how we invent pictorially to create the illusion of reality with light, surface and material in an immersive way are visible in Japanese fusumae (screen paintings), Neo-Assyrian wall reliefs, Roman mosaics and Renaissance frescoes, to name a few (Cohen 2003).

Artists have also embraced lens based technologies to enhance both their perception and understanding of what they see and to work with creatively in constructing their images (Steadman 2001).

There is now considerable interest within the neuroscience community in researching how the brain works when perceiving visual stimuli (Zeki 1999; Purves and Beau Lotto 2003). Visual forms that play with spatial ambiguity may offer insights into the way the brain functions, but artists have known for a long time that we are intrigued by images that appear spatially irresolvable and I am interested in extending this in ways that also play with our preconceptions of how imagery is presented.

While the picture plane may be viewed as a window onto an illusion of space, it is also the surface upon which spatial reality may be rebuilt. In questioning what we observe we must inevitably have recourse to our experience of the physical world around us and the pictorial offers us a way to explore these experiences and reinvent the world we see. Illusion of depth may be conceived as being built upon more than one plane or in such a way that a surface may appear to have many layers of reading. Inventing with the visual topology of the surface, while retaining its flat physical characteristic, has enabled me to create images which alter in their illusions of three-dimensional spatial readings when viewed over time or with varying light conditions and projections.

Fig. 5.1 Nathan Cohen 'Intangible Spaces', RPT Installation, Aisho Miura Gallery Tokyo 2010; computer screen showing format of video for wall projection

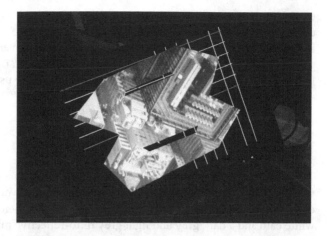

2   Perception and Visual Experimentation: Incorporating Computer Technology in Recent Interactive Artworks

In seeking to invent spatially beyond a conventional pictorial format I have been developing the integration of digital content in artworks incorporating interactive Retro-reflective Projection Technology and real time/time delayed video installations.

I have found that Retro-reflective Projection Technology (RPT) can be used to create visual images on a larger scale with embedded digital projections. By constructing the artwork in particular ways it is also possible to integrate digital content so that it does not appear to be constrained by the fixed framing format of a screen edge.

Tachi Lab (a leading laboratory researching Virtual Reality and Telexistence directed by Professor Susumu Tachi, now located in the Graduate School of Media Design, Keio University, Tokyo, Japan) has developed technology to support research into 'optical camouflage' and 'augmented reality' (Tachi 2010) which I am working with in the development of new visual forms that can also be applied at an architectural scale. This has resulted in the creation of large scale publicly sited artwork installations, displayed at *Ars Electronica*, Linz, Austria (2008), and at the Aisho Miura Gallery in Tokyo, Japan (2010), with ongoing development (see Fig. 5.1).

In creating my artwork I am looking to find a dynamic balance between the real world we live in and the world that may exist within an image and it is in seeking to find ways to link the two that a dialogue between the imagination and reality takes place. The issue of how the brain determines the difference between an object seen and its reflection or likeness is one which has preoccupied those who have studied the nature of visual perception from the neurobiological perspective (Zeki 1999). To find a form of visual expression that relates to the way we see encompassing a panoramic view and at a scale which encourages a more immersive and interactive experience is an exciting proposition. When combined with the possibility of incorporating moving imagery inventive potential is further developed. Enabling the image seen to be interacted with directly by the viewer, even locating their presence within the image itself in real time, confers a further layer of reading and interpretation and this is a process that has been achieved with the use of computers.

3   Interactive Wall Installation

The interactive artwork presented at *Ars Electronica* incorporated three layers of reading with two layers of projection. Initially, the artwork may be viewed in its own right and is composed of irregular four-sided elements made from black and white card and a dark grey and light grey retro-reflective material, all fixed directly onto the wall surface. In normal ambient light conditions this gives the appearance of an image constructed from a specific combination of elements whose tonal values and composition create the illusion of a three-dimensional structure projecting from the wall surface, although it is flat.

When looking through the projector that faces the artwork at a 90 degree angle the viewer initially sees a static structured grid corresponding to the artworks construction, a first layer of projection that reveals more detail about its spatial composition and how this was conceived at the drawing stage.

A motion sensor attached to the front of the projector records with a slight time delay the silhouette of people walking in front of the artwork and this causes a second layer of video projection to be revealed beneath the first layer, in the area rendered visible defined by the edges of the moving silhouette. I had intended that this second layer of video projection would be a view captured in real time of the space behind the wall on which the artwork is installed, further questioning the relationship of the viewer to the seemingly solid surface of the wall on which the artwork is installed and revealing what exists in the space beyond. However, as the view behind the wall was not particularly engaging for the purpose of this exhibition it was necessary to use a video recording I had made earlier in Trafalgar Square, London.

The artwork spanned several meters across a wall and the projector was placed at sufficient distance to allow for the projected images to cover several of the retro-reflective elements composing it, and for observers to walk between the projector and the wall which also triggered the motion sensor.

The projection process has to be constructed with a computer programmed to drive the video sequences and modifying the projections in relation to the sensor input, although the viewer will not necessarily be aware of this as the computer is not visible as part of the display. The results are engaging and stimulated considerable response on the part of the viewers who interacted with the video projection and realized that in the viewing of the artwork they were active participants in the experiences of others who were also observing over time. This occurred in two ways; viewing through the projector as others walked up to and around the artwork, and being observed by others when approaching closer to the artwork.

Fig. 5.2 Nathan Cohen: art
installation, *Ars Electronica*,
Linz, Austria, 2008 (**a**) without projection

The physical properties of the RPT reflective material offer a natural extension
to the media familiar to the painter and allow for pictorial invention at a human
and architectural scale. It can be cut into different shapes, applied directly to and
bent around surfaces of different size and when viewed in ambient light conditions
the light and dark grey of the RPT materials provide a tonal contrast that creates a
balanced scale between black and white. Being highly light reflective when a more
focused light is projected onto their surfaces the tonal scales shift, with the brightest
element becoming the lightest of the grey RPT materials and the white non-RPT
material surface seeming grey by comparison. This offers another means of playing
with perception of space and depth within the artworks' construction and can alter
the illusion of the three dimensionality of the image given varying light conditions
and projection.

In seeking to invent in my own work with images that appear three dimensional
while retaining pictorial form (i.e. they are flat) and that have irregular boundaries I have created structures which are not confined to a picture window illusion
of space. As the image also exists within the 'real' space of the viewer it becomes
more architectonic in its spatial implications and activates a more dynamic relationship with the 'real' space within which it exists. When conceived on a human scale
this allows for the creation of an immersive environment encouraging interaction
with the artwork. When pictorial forms are combined in such a way as to challenge
spatial interpretation a new dynamic unfolds resulting in images which are both intriguing and intellectually engaging. This can be developed overtly or subliminally,
with the structure or physical construction of a work demanding visual dexterity
in its comprehension while the light projected layering of imagery in selected elements offers multiple possibilities for spatial interpretation.

Consequently, I am investigating the potential for spatial invention offered by
the digital projection of images, both still and moving, onto light reflecting material
elements incorporated into my artworks. In their construction play is made between
observing the artwork as a visual construction: a) in its own right without projection; with certain areas exposed to projection, enhancing the three-dimensional
reading of the space within the artwork (see Fig. 5.2); b) optionally a secondary

Fig. 5.3 Nathan Cohen: art installation, *Ars Electronica*, Linz, Austria, 2008 (**b**) detail of projection layer 2

layer of projection embedded 'behind' the first projection layer, revealed through the motion of viewers in front of the projector, all combining to offer extra dimensions to the viewing of the art work (artwork installation at *Ars Electronica*, Linz, 2008) (see Fig. 5.3); and c) the integration of real-time video projection (Aisho Miura Gallery, Tokyo, 2010: floor projection and real-time video wall projection) (see Figs. 5.4, 5.5, and 5.6).

4   Intangible Spaces

Tokyo provided the wider context for the exhibition 'Intangible Spaces' at the Aisho Miura Gallery (2010). The city is a complex and contrasting environment where the new rubs shoulders with the old; buildings constructed using the latest technology can be seen alongside centuries old shrines and temples which are examples of the technological innovations of their day. In the art installations I created for this exhibition I wished to capture something of the essence of this experience, with the gallery itself once being a house used by the local temple, now converted into a modern gallery space. The imagery used in the recorded projections combined two views of scenes observed in Tokyo at the time of the show: for the wall projections I contrasted video sequences of Tokyo panoramas viewed by day from the top of a high building, enabling the complexity of the city to be viewed from an unusual perspective, with the summer *hanabi* fireworks displays, momentary bursts of structured light seen at night.

The artwork was constructed from irregularly shaped and alternating lighter and darker grey RPT material panels located at the corner of two walls resulting in an installation that appeared three dimensional and as if projecting into the viewer's space, onto which the looped video sequence was continuously projected. RPT of-

Fig. 5.4 Nathan Cohen
'Intangible Spaces', RPT
Installation, Aisho Miura
Gallery Tokyo 2010

Fig. 5.5 Nathan Cohen
'Intangible Spaces' (**c**) RPT
floor projection, Aisho Miura
Gallery, Tokyo, 2010

fers the potential for inventing with projection techniques on an architectural scale and my intention was for the distinctions between illusion and the reality of the viewer's space to become blurred, with the artwork to appear to be a part of the real space experience of the viewer.

The design for the artwork was in part determined following initial discussions relating to the technical aspects of the RPT projectors and my decision to include

Fig. 5.6 Nathan Cohen
'Intangible Spaces' (c) still of
real-time video wall projection, Aisho Miura Gallery,
Tokyo, 2010

real-time video in the process, which involved site visits to the gallery with the Tachi Lab team. The first practical stage for installing the artwork involved identify-ing the optimal projection locations within the gallery space, cutting the RPT mate-rial according to the designs and installing them on the floor and walls to allow for the projections to take place. Then the hardware, including 2 RPT projectors, video cameras, computers and cabling were installed. The final stage involved calibrat-ing the equipment and preparing the video sequences I had shot earlier for the wall projection.

The RPT projectors had to be located at specific distances from the wall and floor and were oriented so that the viewer could easily navigate the space and view the artworks both with and without the projections. The floor piece incorporated a real-time video projection of the view as if seen directly through the floor to the room below, with the video camera located at an equivalent angle to the projector. This gave the viewer the impression they were seeing through the floor to the activ-ity in the space below.

Both the floor and wall based RPT artworks required interaction on the part of the viewer to fully observe their pictorial potential, for while the installation in-cluded the wall and floor pieces that could be viewed as images that interacted with the architecture of the gallery space to create the illusion of three dimensional struc-tures, when the projections were also seen through the RPT projectors an alternative reading of the space within the artworks was enabled.

This often resulted in viewers spending some time walking up to the artworks themselves, then back to viewing them through the projectors, changing their view to look again without projection, and so on, encouraging a high level of engagement in encountering and looking at the installation. The intention was also to challenge pictorial convention in the way the artworks were located in space, utilizing the corner of the room and floor, locations not usually associated with the display of pictorial images.

5   Encountering Ourselves in the Interpretation of an Image

We are very familiar with looking into an imaginary space, and developments in immersive technologies have sought to place the viewer in ever closer proximity to the action. A third artwork displayed in the Tokyo exhibition develops the idea of locating the viewer within the space of the artwork itself and utilizes an approach to visual invention which could only be achieved using the computer.

As the visitor to the exhibition climbed the stairs to the installation located on the second floor their progress was being recorded by a small video camera. A short time lapse of the resulting video image was combined with isolating the visitor's image from the background and a moving video sequence I had made of the art installation itself. The result was a video projection, which the visitor could view as they reached the top of the stairs, of themselves 'climbing into' and located within the artwork they would not yet have seen. This formed the start of the interactive process of viewing the art installations.

Keitaro Shimizu, who assisted with the installation projections, describes the computer software used to achieve this: 'We could make this projection by use of a computer program utilizing 'difference of background', enabling differentiation between two pictures, allowing us in this case to determine which object is moving. Using this program we initially recorded the background image of the view of the stairwell with no motion which enabled us to compare the color of the real-time moving image, subsequently captured by video camera, with the color of the background image. Where a significant difference between the colors of two images occurs for the purposes of this display we can recognize it as motion, isolating the pixels whose colors have changed and combining this image with the background image. For this projection a slight a time delay of a few seconds was also incorporated enabling the viewer to see their progress up the stairs as they emerged into the gallery space art installation. After this the program resumed recording mode; if no motion was detected the projection replayed the last motion sequence until new motion was detected which then started a new cycle of recording and projection.'

Memory forms an important part in our interaction with the world around us. There are many ways in which this can be explored and artists have sought to create images and art forms that rely for their interpretation upon experiences we have gained prior to their viewing. I am interested in creating artworks that encourage the viewer to be an active participant in the experience of encountering what they see. In this artwork this relies on a technological application that combines a small time shift and its projection into the moment of engagement—the person climbing the stairs encounters themselves in this act as they arrive at the point of viewing the artwork for the first time, and this marks the initial moment of engagement with the exhibition.

The seemingly mundane process of walking up some stairs can be represented to the viewer in a form that is not conventional, but relies for its interpretation on understanding certain conventions relating to how we interpret what we see. The context for this plays a part in how we choose to interpret what is unfolding before us, and the viewing of the artwork in the specific space and context of the other artworks that are also displayed in this video sequence requires the viewer to engage with the totality of the experience to be able to comprehend what is seen. This is a process that unfolds over the time it takes to view all the artworks in the exhibition, and rewards a return to the initial video sequence as, having seen the exhibition, the viewer can now understand the context for the imagery they initially see themselves ascending into in the time delayed video sequence.

6 Back to the Computer Screen

Up to this point I have been discussing the application of the computer in the context of projected imagery within an architectural space. The interactive experiences described in the artworks exhibited at *Ars Electronica* and Tokyo pose an interesting challenge in recording and documenting them as they are temporary installations. In considering this I have also been exploring the creative potential of the space offered by the computer screen and this has led to new artworks that engage the viewer dynamically in the act of viewing on a screen.

We are familiar with, and would normally expect, that an image fills a screen. We are also familiar with more than one image being presented on a screen simultaneously, most typically with the multiple screens replicating the rectilinear form of the screen itself—windows within windows. The interactive nature of computer technology combined with the limitation of the flat screen encourages us to layer our experiences, moving from one window to another often displayed simultaneously.

In considering how pictorial form can express different types of interaction with space it is necessary to look at the relationship of edge to surface, for the picture plane is often clearly defined by boundaries which separate it from the non-pictorial space of its surroundings, particularly on a more intimate scale.

It is also important to think about how we perceive moving images; the saccade sequences of moments in movement neurologically processed giving the sense that these are contiguous over time. Then there is also the question of focus and angle of vision to be considered—how close or distant we are from what we see, and how we are located in relation to an action or object of interest and how this might alter our understanding of it. Light conditions also vary our perception of space.

I have been experimenting with animated sequences of recorded video images that are combined and composed over time and are formally articulated on the computer screen in ways that shift how we perceive them spatially and inform our memory of the actions that occur.

Recent test pieces introduce fragmented and multi-faceted views of video sequences of human actions, such as walking through a space or cooking a meal. The shifting composition of these facets is intended to create a dynamic tension between the formal elements that compose an image and the edge of the space proscribed by the screen. With several views of the same activity displayed simultaneously

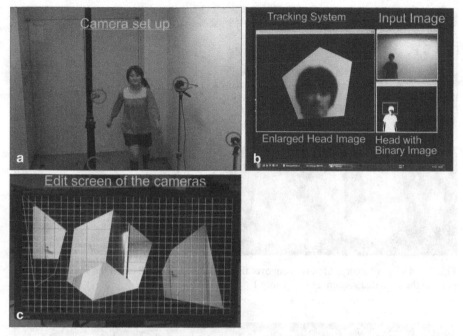

Fig. 5.7 Images of cameras and computer set up

within animated facets that shift over time it is possible to impart a sense of how it might feel to be engaging in the activity being viewed and to see this from multiple perspectives enhancing a sense of depth and movement within the picture plane of the screen and creating a more immediate presence of the activity. To this end I have found that by liberating the pictorial elements that compose this constantly changing sequence of images from the rectilinearly defined edges of the screen space it is possible to create a more dynamic sense of space, time and movement.

One example of the set up for generating these multiple images that compose these video animations consists of a group of video cameras placed in different locations along the path a person walks (see Figs. 5.7 and 5.8). Each camera relays its input to a computer that has been programmed to generate a sequence of shaped facets that move and combine across the picture plane of the computer screen and which incorporates the different video camera images. This is an artwork that can be viewed in real time or as a recorded sequence, the former offering an interactive element that allows for the viewer also to become the subject of the recordings if they chose to walk through the cameras.

In another artwork a similar approach to multiple cameras and viewing angles is employed with the intention of recording a sequence of events over a period of time (see Fig. 5.9). Here I have composed an animated video sequence that conveys the experience of preparing and eating a meal. I have also extended the moving facets to meet the edges of the screen actively incorporating them into the spatial construction as each of the facets alter their position relative to one another in

Fig. 5.8 4 screen captures of movement over time (each image is a view on the screen at a given point in the animated sequence)—sequence 1

Fig. 5.9 4 screen captures of movement over time (each image is a view on the screen at a given point in the animated sequence)—sequence 2

the unfolding sequence, rendering the screens edges a dynamic parts of the video imagery.

Different degrees of focus are also utilized between the cameras offering detailed and broader angles of sight. This is significant, for in our interaction with the environment we are constantly altering our point of view in order to negotiate and better comprehend what it is we are looking at and engaging with. (Henderson 2013) It is in the editing of these multiple strands of video in conjunction with the animated facets that the artwork takes form.

As an artist I am seeking to interpret the nature of what I see through the artworks that I make and in the process encourage others to see the world afresh and perhaps in new and different ways. We are now living in an age where digital media offers for an artist another medium for visual invention, and in my utilization of computer articulated images in these artworks I am doing so with the purpose of enabling and enhancing perception. As Professor Susumu Tachi observes: 'The word 'art', which currently refers mostly to fine art, also meant 'technology' in ancient Greece. In other words, art refers to both fine art and technology and it originally meant the act of creating something new.'

References

Cohen, N. 2003 'Illuminating Thought: Imaging Reality' http://www.nathancohen.co.uk/Nathan_Cohen_Web/writing/Nathan_Cohen/pages/page1.htm

Henderson, L.D 2013 'The Fourth Dimension and Non-Euclidean Geometry in Modern Art' Leonardo Book Series

Kemp, M. 1990 'The Science of Art: Optical Themes in Western Art from Brunelleschi to Seurat' Yale University Press

Purves, D. and Beau Lotto, R. 2003 'Why We See What We Do' Sinauer Associates Inc.

Steadman, P. 2001 'Vermeer's Camera' Oxford University Press

Tachi, S. 2010 'Telexistence' World Scientific Publishing Company

Zeki, S. 1999 'Inner Vision' Oxford University Press

Chapter 6
Wasting Time? Art, Science and New Experience. Examining the Artwork, Knowmore (House of Commons)

Keith Armstrong

1 Introduction

Today the future is travelling rapidly towards us, shaped by all that which we have historically thrown into it. Much of what we have designed for our world over the ages, and much of what we continue to embrace in the pursuit of mainstream economic, cultural and social imperatives, embodies unacknowledged 'time debts'. Every decision we make today has the potential to 'give time to', or take 'time away' from that future. This idea that 'everything' inherently embodies 'future time left' is underlined by design futurist Tony Fry when he describes how we so often 'waste' or 'take away' 'future time'. "In our endeavors to sustain ourselves in the short term we collectively act in destructive ways towards the very things we and all other beings fundamentally depend upon" (Fig. 6.1).[1]

Economics, science, technology and commerce are routinely painted as the fundamental creators and drivers of our future possibilities, whilst the need for fundamental cultural or political shifts are much less often factored into this equation. However a subset of cultural theorists, activists, artists and futurists have begun to illuminate the urgency of embracing fundamental cultural, and consequent behavioral changes in order to devise transitional pathways towards sustainable futures. Their thinking goes far beyond the often shallow 'greening' of business, architecture, consumption and culture, instead suggesting a project that lies far beyond much of today's popular imagination. The magnitude of this idea is encapsulated by design futurist thinking from those such as Tonkinwise, Ann-Marie Willis and Manzini, and particularly Tony Fry in his notion of "The Sustainment"—something he describes as, if implemented, the largest social, political and environmental shift in thinking and action humanity would have experienced since the Enlightenment.

---

[1] Fry 2009.

K. Armstrong (✉)
QUT Creative Industries, Brisbane, Australia
e-mail: k.armstrong@qut.edu.au; keith@embodiedmedia.com

N. Lee (ed.), *Digital Da Vinci,* DOI 10.1007/978-1-4939-0965-0_6,
© Springer Science+Business Media New York 2014

Fig. 6.1 Interacting with Knowmore (House of Commons). 2009, Image Sonja de Sterke

The Sustainment is a very big idea, a mind blowingly big idea. It is an idea that leaves current thinking of sustainability, and the likes of 'natural capital' and 'triple bottom line reporting' on the shelf and in the shade.

It is an idea we need to creep up on.

Our starting point is to recognise that the idea of sustainability is lodged in a limited and now largely debased agenda. It's about propping up the status quo rather than making the means of redirection towards viable futures. De facto, much 'sustainable architecture' and many 'sustainable products' are implicated in sustaining the unsustainable. Equally, 'sustainable development' is bonded to 'development logic'—the 'logic' of continual economic growth—rather than the development of sustainment. It does not add up to the fundamental directional changes essential if the human race is to stay around.[2]

Projects such as 'The Sustainment' propose a root and branch re-designing of how we think about ourselves, and therefore consequently how we might then act in the worlds that we create; which themselves form within the greater world upon which we all depend. Change on such a scale of reach and complexity is as unimaginable for us today as would have been the changes coming over the horizon for pre-Enlightenment society. Such a wholesale 'ontological re-designing' suggests that we need to become in essence different kinds of human, propelled by alternate desires and with quite different understandings of what constitutes progress. Far-reaching visions such as these will always require the visionary thought of a relative few initially, accompanied by the production of new kinds of powerful images and experiences that might then help propel the broad scale take up of these ideas into the longer term future. It follows therefore that such change will unlikely be embodied simply through the logical processing of information, but will ultimately require a fundamental shift in 'hearts and minds', something that is arguably a central part of the work that many cultural practitioners already do (Fig. 6.2).

---

[2] Fry 2006.

Fig. 6.2 The author interacts with Knowmore (House of Commons). 2009, Image Sonja de Sterke

Visions of this magnitude, and the projects that surround them, must by definition sit at the nexus of science and culture—a truly potent place where urgent conversations are beginning to form and shape.

It is these nascent understandings, and my own compelling desire to make whatever contributions are possible to this meta project, that has long driven my practice. I will now further frame this 'politics' and outline my artistic response via a resulting media artwork called *Knowmore (House of Commons)*.[3] I will also describe the relationship between the ideas and thinking that both instigated it and underpinned its development and presentation, and explains how they ultimately manifested in the final work. I will then discuss how I use such work as a conversational vehicle or point of instigation to discuss the ideas put forward by Fry, Tonkinwise, Ann-Marie Willis and others.

2   Wasting Time

We all make inadvertently 'time-wasting' decisions on a daily basis; decisions that frame in some way how the future will then manifest for others. We do this in mostly small and seemingly insignificant ways, and in virtually every case we are, and will probably never be, called to account for our decisions. If we were however able to plot the relational outcomes of our daily actions into the future (e.g. with respect to our consumption, travel or leisure), we would undoubtedly pause to reconsider the tacit assumptions upon which those daily decisions are made. Necessarily living would become far more complex. Take the thorny example of air travel for example. We do have some increasing awareness about the time debt incurred by jet emissions, but we have in no way reconciled how we will deal with that as a global community. Every time we jump on an aeroplane and cross the globe we

---

[3] Armstrong et al. 2009.

Fig. 6.3 Screen capture,
Knowmore (House of Com-
mons). 2009, Image Sonja de
Sterke

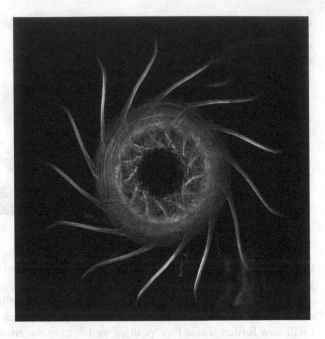

contribute in some way to future climatic chaos.[4] Nonetheless, this inconvenient
fact hasn't significantly mitigated flight frequencies. For those of us who choose to
think through such dilemmas, the environmental, social and political costs of something like mass fuel consumption has to be somehow subjectively balanced against
the social, cultural or political benefits of making such journeys—a particularly
DIY process of (uncertain) self-justification. Set within a general climate of peer
inaction or indifference there are few if any fundamental tools or resources that can
really help us to act upon these facts at more than a cursory level (Fig. 6.3).

Of course a few of us may choose to purchase carbon credits, or skip or combine a trip or two, but beyond those small tokens we all typically join the airborne
throngs. Whilst we 'know' the stakes through freely available information, the ultimate affect on others, the world and a future seem understandably abstract and
far away. The problem is most often seen as being 'out there', sitting, awaiting the
technical expertise of others to solve.

But do we really have this luxury? Do we fundamentally misunderstand the time
our species has left so badly? Even if there is still no broadly understood and applied framework on what sustaining the future requires, the ecological crisis we
are facing cannot sensibly be conceived as being simply 'out there'. Nor can it be
conceived of as just a series of technical problems waiting to be solved, because in
reality the crisis is in here: We are the crisis.

---

[4] Monbiot 1999.

3   Cultural Change

Despite our access to pertinent knowledge increasing exponentially we are increasingly becoming a threatened species, because we are not fundamentally learning and adapting to the uncomfortable facts of climate chaos, environmental degradation and cultural stasis. Most of us would see the root and branch change those such as Fry and others have proposed as both inconceivable, too hard or simply futile—in other words 'a waste of time' (right now)—and yet the irony remains that by not at least attempting to project our actions into the future as a means for refining them, we are blatantly 'wasting time' for those who come after us.

And so the designs that scientists, economists, miners, manufacturers, artists, accountants, food manufacturers, farmers, product designers, technologists and builders produce for us, and that go on themselves designing through their usage, collectively define this shrinking of the future, thereby refining what will and will not be possible for the future.

The future is of course not an empty vessel waiting to be filled by our innovations. The future is already a place replete with the designing power of things, processes and structures from the past. How could, for instance, Henry Ford have realised how his groundbreaking designs would go on to fundamentally re-design the city, architecture or the atmosphere? How could those who first mined coal to tap its rich, embodied energy have known or begun to predict that it would result in the climatic instability that we are witnessing today. Once we are weaned onto practices such as fossil fuel incineration, how can we then be triggered to seek alternate paths whilst the resource remains; paths that better protect, rather than wilfully destroying the future (Fig. 6.4)?

4   Lost Time

Writing in 1978, Bernd Magnus[5] profiled a key idea that he named 'kronophobia', describing how we are simultaneously both fearful and ignorant of the nature of time—particularly long time periods as opposed to the short snippets that we use to frame our life and work processes. Through his considerations of Friedrich Nietzsche's writings he analyzed how we have long invested ourselves in furthering the illusion of human permanence.

Magnus recognized that what Nietzsche made clear is that 'we' seek permanence where and when there is none. Moreover, our very being, our being towards death, is enacted via refusal of time, as the pursuit of power, wealth and fame evidence. To make time we have to understand time not as measure but as change, with everything having its own time.[6]

---

[5] Magnus 1978.

[6] Fry 2010.

Fig. 6.4 Screen capture, Knowmore (House of Commons). 2009, Image Sonja de Sterke

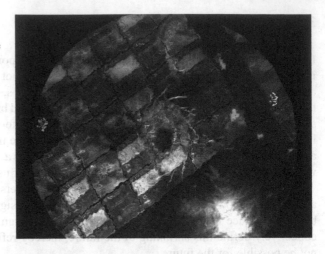

This suggests that time should be better thought of as a kind of 'medium', recognised for its many different scales and cycles—for example the time of our lives, the time of old age, the time of mammalian breeding, deep geological time, planetary formation timescales and so forth: all constituents of a kind of meta-time that lies far beyond conventional conception.

The virtually unequivocal voice of the community of climate scientists, the IPCC[7] is probably the most pointed example of both the outcome of this misunderstanding. We know that our long term and ever increasing appetite for converting stored carbon into carbon dioxide present an increasingly significant time debt to be shouldered for those who come after—lessening the time that these future generations will be able to enjoy the relative climatic stability that we have enjoyed in our short time on this planet. Here time in all of its dimensions has become 'finitude' and slowing this king tide has become an extraordinarily complex, shifting problem that challenges us to our ontological core.

However, understanding that 'everything has its time' is also an unexpectedly powerful thought that might allow us to better frame our journey towards future change.

5   Knowmore (House of Commons)

Set within this thinking, I now will examine the motivations behind a major media artwork *Knowmore (House of Commons)*, (first shown at the State Library of Queensland in 2009 and then in the Mediations Biennale, Poznan Poland in 2010); an embodied, interactive installation investigating the cultural dimensions of sustainability and time (See Figs. 6.1, 6.2, 6.3, 6.4, 6.5, 6.6, 6.7, 6.8, 6.9, 6.10, 6.11 and 6.12).

---

[7] Secretariat 2013.

Fig. 6.5 Screen capture, Knowmore (House of Commons). 2009, Image Sonja de Sterke

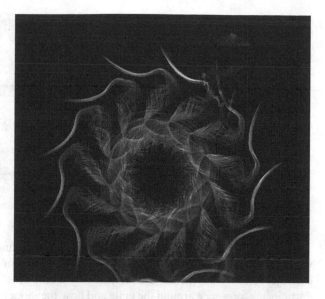

Fig. 6.6 The author interacts with Knowmore (House of Commons). 2009, Image Sonja de Sterke

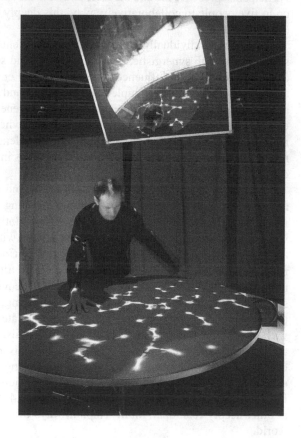

Fig. 6.7 Screen capture, Knowmore (House of Commons). 2009, Image Sonja de Sterke

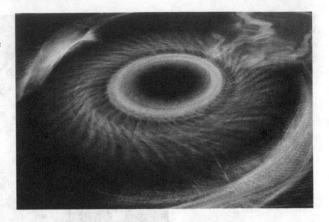

A large circular table spun by hand (see Fig. 6.2) and a computer-controlled video projection falls on its top (see Fig. 6.4), creating an uncanny blend of physical object and virtual media, accompanied by a real-time, six-channel audio work. Participants' presence around the table and how they touch it is registered, allowing up to five people to collaboratively 'play' this deeply immersive audiovisual work (See Fig. 6.1).

Participants individually and collectively experience a range of time- suggestive scenarios through synergistic, generative image and sound, allowing them to both directly and indirectly influence a complex digital environment that computationally mimics the temporal complexities of natural and artificial systems. This hints at broader ecological/cultural concerns and more generally relational timescales by encouraging each participant to look for moments where small strategic actions can make significant contributions to the whole, challenging them to image how we might "go beyond": i.e. take newly conceived steps in a collective consideration of our temporal futures (Fig. 6.5).

In this way the work also subtly asks what kind of resources and knowledge might be necessary to move us past simply knowing what needs to be changed to instead actually embodying that change; whilst hinting at other deeply relational ways of understanding and knowing the world. Set within an age in which we arguably 'misunderstand' time, *Knowmore (House of Commons)* considers the urgent need for us to better celebrate and 'care for' those 'times' which we share in common: scales that mark the cultural and biophysical environments that fundamentally sustain all life today. This requisite shift in thinking, action and knowing suggest that we need each envision new ways to re-orientate our everyday life choices in ways that better respect those commonalities, whilst also respecting the differing times of each and everything. It further suggests this idea by focusing on the power of embodied learning implied by the works' strongly physical interface (i.e. the spinning of a full size table) (See Fig. 6.6) alongside the complex field of layered imagery appearing upon that table top (See Figs. 6.4, 6.5, 6.7 and 6.11) which hints at other deeply relational, multi-temporal ways of understanding and knowing the world.

Fig. 6.8 Knowmore (House of Commons) presented at the State Library of Queensland. 2009, Image Sonja de Sterke

Fig. 6.9 Screen design based on the periodic table, Knowmore (House of Commons). 2009, Image Sonja de Sterke

Fig. 6.10 Screen designs based on the periodic table, Knowmore (House of Commons). 2009, Image Sonja de Sterke

In this respect *Knowmore (House of Commons)* revolves around Magnus' notion that everything has its time—a key idea that begins to make fuller sense when one turns for a moment away from the predominance of time solely conceived of as a linear arrow that typically guides our daily actions and thoughts.

Fig. 6.11 Interacting with
Knowmore (House of Commons). 2009, (Image Keith
Armstrong)

Fig. 6.12 Elements from screen designs based on the periodic table, Knowmore (House of Commons). 2009, Image Sonja de Sterke

Set within this temporal space *Knowmore (House of Commons)* is envisaged as a speculative hybrid of art and design thinking and practice that lures its participants into a consideration of personal and collective journeys of 'futuring'. *Knowmore (House of Commons)* therefore asks us to picture futures not as voids awaiting input, but rather as fulsome storms: replete with much of the detritus of historical decision. In this way it seeks to challenge participants to think of themselves more as pathfinders rather than pioneers, clearing space for contemplation and potential futural actions. Seen in these ways it acts as a complex hybrid of participative process, contribution and visualization: as well as an actively offered invitation to re-think much of what is routinely presented to us as given.

And so, this dilemma of time as a medium (rather than machine) becomes the central conceptual thread of the work (Fig. 6.6).

6 Commonality in Difference

Tony Fry describes how creating a new reality of future-making will require both ontological and consequent political shifts that will deeply challenge the status quo of democratic politics—aligned as it is not with the extending of time left, but typically with shorter term and often time-subtracting agendas. Critically Fry calls for place of the creative practitioner to be recognized if the kind of sociocultural political imaginaries implied by 'The Sustainment' project are ever to be realized. His strategy therefore necessarily supports the active evolution of new forms of design and designing practices that function beyond what Bataille called the "restrictive economics" of capitalism and the highly aligned cultures that predominate contemporary, globalised life. Fry calls for no less than a wholesale transformation of our being by design.

Restating design's endless circling: anthropologically, as designers we are equally the designed. We come into being by design as much as we arrive biologically and socially formed. I go so far as to say Darwin and Herbert Spencer were blind to the directive force of technics and the artificial.[6]

Knowmore (House of Commons) aligns itself with Fry's expressed need to expose the broader agency of creative practices within a personal micro-politics of change making. I carry this spirit and purpose into my own long standing practices within the realms of media arts and art-science collaborations, fertilized through many conversations and workshops I've attended over the years with both Fry and his colleagues. Both Fry's and my own intentions broadly converge in a shared community of interest and intention, although we also each understand that our commonality in difference then allows the fruitful and fluid transference of many of the foregrounding ideas to series of questions take their own routes when applied into public presentational forms (Fig. 6.7).

7 My Own Journey

This challenge entails and necessitates us all to each think about moving exiting cultural disciplines beyond the limits of how they are currently and conventionally understood and practiced. This stands as my own driving aim, and *Knowmore (House of Commons)* is but one thread of that time-infused journey. Towards these ends, the loose collaborative organization that I direct, *Embodiedmedia*[8], has long pursued such goals by representing a multidisciplinary team of collaborators and researchers who are collectively motivated to think about how we can each best create sustainable pathways to the future—acknowledging that we live in an era that is in many ways 'Post Natural'. Our modality of participative practice is described by Pat Hoffie[9] as having a focus upon "deep collaboration in the process" and making that also "invites collaboration as an integral aspect of experiencing it".

We achieve our aim through engaging the public in deeply considered artistic works of all kinds, working in ways that avoid streams of facts or stern lectures; working instead to influence mind, body, emotion and scientific process in all areas of society and culture.

I personally have been a practicing freelance artist, creative director, media designer & system integrator of new media artworks since 1993, specializing in the development of collaborative, mixed reality productions that merge site-specific interactive installation, performance and multimedia practices. These art works include site-specific electronic arts, networked interactive installations, alternative interfaces, performance forms, public arts practices and art-science collaborations. Overall my research and writings focus on better understanding how scientific and philosophical ecologies can be used to influence and direct the design and conception of new artworks (Fig. 6.8).

8 Time Manifesting

The first iteration of *Knowmore (House of Commons)* was presented in State Library of Queensland in 2009 (See Fig. 6.8) as part of a residency project for which I had been commissioned, and the second was accommodated in the darkened room of a medieval castle for The *Mediations Biennial of Art* in Poland in 2010. In the former setup you brushed past a circular curtain and entered into a dark, private space. In both setups, either with others or on your own, you were subsequently presented with a round table, upon the face of which graphics were seamlessly integrated with the tabletop whilst a responsive multichannel spatial soundscape further animated the space. Whilst the amoeba-like visuals moved gently and the sound was audible the work only began to evolve into coordinated motion and volume once one or

---

[8] Armstrong 2013.
[9] Hoffie 2009.

more people grabbed the edge of the table and spun it clockwise or anti-clockwise directions. The three modalities of interaction were spin speed, physical location around table and surface touch with the work being built from a range of different scenes that merged seamlessly with a further deeper phase descending the user.

Part of the user experience is described by reviewer Greg Hooper.[10]

The graphics are great—first up I'm seeing some acid green polyps, chasing each other around and around the table, flagella beating away behind them, tentacles fluffing about in front. They twist and turn, and the faster the table spins the more it acts like a centrifuge, driving the polyps outward to the edge.

There is a genuine sense of watching something alive and swimming in a current controlled by the lazy sense of watching something alive and swimming in a current controlled by the Lazy Susan spin of the table.

Another scene and another biomorphic form—circular, symmetric, hairy with cilia, more varied in colour than the polyps. I spin faster and the image speeds up and zooms to cover all the table surface. The physics of the animation is great—utterly natural in the way the cilial hairs are pulled out straighter and straighter as the speed increases. But it is the clarity of the interaction that stands out. There is no apparent lag between giving the table a spin and seeing changes in the animation— the table is beautifully engineered and the interface between table and animation is completely transparent.

Rather than imitating life forms, the next two animations show particles and clouded shapes sucked down into the centre, faster and faster with the table spin.

Furthermore Antoanetta Ivanova writes in a pre-review article[11] for *Impact '09*, (Staged at the time of the Copenhagen Climate Talks),

The work computationally mimics the complexities of natural and artificial systems, which do not follow linear principles but are 'composed of multiple series of parallel processes, simultaneous emergences, discontinuations… and mutations of every variety'. Thus, through the interactivity we are asked to share mutually sustaining systems and are encouraged to look for particular places in them where our small strategic actions could pay off in big results. The work draws parallel with the processes of the public governance of climate change that are self-organised, interlinked and bottom-up. The more we interact, the more we know what it is that we need to do—individually and together—in this seemingly anarchic system (Fig. 6.9).

Whilst components of the work implicitly support the idea of the time of things that are other, strange and alive, beneath this model sits another very human orientation of things, a matrixical structure that was introduced into the artwork to speak about more about our reliance on entities rather than relationships. This is invoked by an old fashioned rendering of a periodic table but with the elements and texts reorganised into imaginary forms, that whilst apparently discrete, each invoke leading quotes from Tony Fry and others on the notions of relational time and ontological designing, as well as texts and images that hint around sustaining and non sustaining conceptions of world (Fig. 6.10).

---

[10] Hooper 2009.

[11] Ivanova 2009.

Greg Hooper discuses this temporal component in his review[10]:

Moving around the table a little more and we see a faded database on old style microfiche—blue grey images and defocused text. Now the table spin moves our view on to the image, like driving crosshairs with a shuttle wheel. Spin one way for the X- axis, reverse the spin for Y. There's a nostalgia evoked by the monochromatic microfiche that is reinforced by arranging the information into cells. It's like the periodic table of elements with micrographs and quasi-elementary chemical symbols: Er for Erbium, and Ke for... what? But the nostalgia is not just for a look, but for a system of knowledge based on recording, codifying, naming—partitioning the phenomenal world into atomic events (Fig. 6.11).

I play some more, spinning and slowing, trying to make out the images, reading the grainy cells where I can: "Observation 03: the action of language and knowing are of different registers"; "Observation 06: the should of conduct exists in relation to a crisis in conversations rather than an embodiment of concern."

As I read the information in the database cells the animation slows, the direction changes. Text and image ripple like flowing water, then submerge completely. A few more steps around the table and the polyps emerge again—a few living cells chasing each other, as at the beginning (Fig. 6.12).

9   Concluding Thoughts

The emphasis in *Knowmore (House of Commons)* therefore lies for the participant ultimately in the temporal connectivity and inseparability of embodied experiences, which through their improvisational actions indirectly prompt them to taste the work's underpinning ideas, enhanced by their implicate part within the work's evolving audiovisual imagery. *Knowmore (House of Commons)* therefore offers audiences powerful images suggestive of transformational potential whilst also imprinting an associated sensibility and purpose through embodied, exploratory experience. These inclusive strategies avoid a simplistic reliance on the promotion of fear and guilt around the meta issues of time and sustainability, in the assumption that contemporary audiences may well already be largely inured to warnings about our deepening ecological crisis. Furthermore it introduces complexity through apparent simplicity.

Ultimately my hope is that this combination might in some small way encourage an increasing chorus of calls towards personal engagement and action.

It would of course be simplistic to suggest that participating within such an experience might somehow lead to change in future behaviour or action through some 'transcendental' moment. However it is ultimately the possibility of catalytic reaction in participating audiences, inspired in part by the experience of creative work such as this, and set in the context of all other experience, that as both an artist and a social activist inspires me to continue to create these types works over the decades.

This approach, I hope, ultimately serves to lessen the risk that the real problem of time will remain somehow concealed—i.e. US.

10   Credits

Dr. Keith Armstrong (Artistic Director) worked in close collaboration with Dr. Chris Barker (3D Visual Design), Darren Pack (3D Authoring), Luke Lickfold (Sound Design) and Stu Lawson (3D Design). The project has been created with Artworkers Alliance and supported by The Australia Council, Arts Queensland, QUT Creative Industries and e2evisuals.

References

Armstrong K. Embodiedmedia: Armstrong, Keith, 2013.
Armstrong K., Barker C., Pack D., et al. Knowmore (House of Commons). Brisbane, Australia: State Library of Queensland, 2009; A 3D, Interactive Table-Top Based Installation.
Fry T. Introducing the Sustainment,. 2050 'Building the Future'. RMIT, Melbourne: RMIT, 2006.
Fry T. Design Futuring: Sustainability, Ethics and New Practice. New York: Berg, 2009.
Fry T. Time and the Political. Mediating Practices: Design, Politics and their Publics Leverhulme Goldsmiths Media Research Centre, University of London 2010.
Hoffie P. Spin-Doctor (some notes on KNOWMORE [House of Commons]). Brisbane: State Library of Queensland, 2009.
Hooper G. Animating the Interactive Spin. RealTime, 2009;34.
Ivanova, Antoanetta, Impact By Degrees, in The Danish Journal of Contemporary Art, 2009.
Magnus B. Nietzsche's Existential Imperative (Studies in Phenomenology and Existential Philosophy) Bloomington and London: Indiana University Press, 1978.
Monbiot G. Flying is Dying. The Guardian Onine, 1999.
Secretariat I. World Meteorological Organization. Geneva: Intergovernmental Panel on Climate Change, 2013.

Chapter 7
The Information Train

Diomidis Spinellis

1 Introduction

When I was a kid I had a pretty good idea of how most appliances in our home worked. The phone was a circuit that physically connected the microphone at each end with the speaker at the other end. The record player's needle picked up the sound from the grooves of a rotating disc and converted it to current through a magnetic coil. Even the TV was a relatively simple affair: two electromagnets had a ray scan the picture while it was modulated to turn parts of the screen white.

How things change… Nowadays to give a similarly realistic picture to our generation's children I have to talk about analog to digital conversion, CPUs, flash memory, compression, psychoacoustic coding, packet routing, pixels, color perception, and liquid crystal displays. Or lie.

Yet, there is still value in understanding the basic principles of modern communication technologies, even at the cost of brave simplifications, for this will seed in our children the interest in the world surrounding them, the willingness to explore it, and, maybe, the ambition to pursue a personally and socially rewarding career in science and engineering.

In response to this goal, I created a scientific experiment exhibit that physically demonstrates how computers communicate with each other by setting up a network in which a model Lego train transfers a picture's pixels from one computer to the other (Fig. 7.1). In brief, the sending end computer (on the figure's left) scans a simple picture from left to right and from top to bottom, and directs a model train (on the front) to send that pixel to the receiving end computer (on the right). This is done by sensing the approaching train and switching a rail junction (front-left) depending on whether a pixel is on or off. The train carries on its top a horizontally-mounted L-shaped piece, hinged in a way that allows it to rotate so that it protrudes

D. Spinellis (✉)
Department of Management Science and Technology,
Athens University of Economics and Business,
Patision 76, 104 34 Athens, Greece
e-mail: dds@aueb.gr

N. Lee (ed.), *Digital Da Vinci*, DOI 10.1007/978-1-4939-0965-0_7,

Fig. 7.1 Experimental setup

from the train's left or right side. This part carries the data between the two computers. Depending on the track to which the junction sends the train, the train will pass close to a column on its left (red) or right hand side (blue), thereby rotating the L-piece to the corresponding direction. At the receiving end, a sensor (yellow, on the right) detects the train's passing, and a second one (green-blue, on the front right) checks to see the shape's orientation. Based on that input the receiving-end computer reassembles the picture bit-by-bit, pixel-by-pixel.

The exhibit is based on readily available components. The use of a large-scale (Duplo) Lego train provides a robust, accessible, and configurable platform to which children can easily relate to. The sending and receiving computers are laptops. In contrast to desktop computers, these are self-contained, and can therefore easily communicate their sending or receiving function by placing them near the positions of the track related to it. One of the laptops is a One Laptop Per Child (Lee 2006) XO-1 model, thus demonstrating the machine's educational potential. The interfacing parts are bespoke circuits based on cheap electronic components. Although the same functionality could have been achieved using Lego-provided black-box components, like WeDo or Mindstorms, the chosen alternative is more open and affordable. Most modern PCs lack simple general purpose input output ports. In particular, USB ports require complex interfacing hardware and device drivers. I therefore repurposed ports provided for other purposes. Nevertheless, the USB ports proved useful for providing a relatively clean 5V supply needed to power the sensors.

2 Sender Implementation

On the sender side (Fig. 7.2), interfacing takes place through the laptop's parallel printer port. This supports eight output bits and five input bits used for signaling conditions like "printer busy" or "out of paper". A motor is used to control the rail junction's position through a rack and pinion assembly. The motor control circuit

(Fig. 7.3) uses two of the printer port's output bits to control the junction's motor, according to the following table.

Bit A	Bit B	Motor
0	0	Stopped
0	1	Forward
1	0	Reverse
1	1	Not allowed

Fig. 7.2 The sending computer

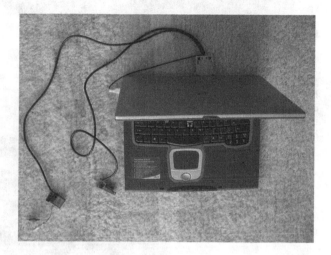

Fig. 7.3 Sender motor control circuit

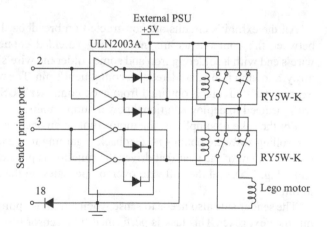

This is done by using a Darlington transistor array integrated circuit (Texas Instruments 1976) to control the windings of two relays, which in turn switch externally-supplied current to the motor. One of the relays controls the motor's power and the other its rotational direction. The Darlington transistors were wired in pairs to increase their current driving capability.

Fig. 7.4 Sender motor control physical construction

Fig. 7.5 Junction driver
construction. On each end of
the rack are the constraining
brakes.

All the exhibit's circuits are constructed on breadboard (Fig. 7.4). Connections
between the components are made using shielded copper wire. By heating the
wire's end with a soldering iron and some solder the wire's enamel insulation melts
away, allowing it to be soldered to a component's pin. The power required for all the
experiment's circuits is obtained from each computer's USB port, thus doing away
with the need to deploy additional power supply units.

For the sake of simplicity, the motor runs in an open-loop configuration, i.e. the
controlling software obtains no feedback regarding the junction's position. Instead,
the movement of the gear that drives the function is physically constrained on both
ends (Fig. 7.5), and the software overcompensates on the time it allows the motor
to run.

The sender side also needs to sense when a train is approaching in order to transmit the next pixel. This task is performed by a sensor board mounted vertically by
the side of the rails (Fig. 7.6). The sensor circuit (Fig. 7.7) is based on an integrated
reflective optical sensor with transistor output (Vishay 2012). This combines in a
single package a 950 nm infrared emitter and a matching phototransistor. The package's window contains a daylight blocking filter, thus improving the construction's
noise immunity. The transistor's output is fed to a voltage comparator designed to

Fig. 7.6 Train sensor board and its mounting

Fig. 7.7 Train sensor board circuit

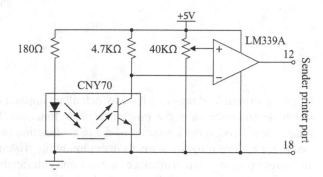

operate from a single power supply (Fairchild 2012). This is used to convert the phototransistor's varying output voltage into a TTL-compatible digital signal that can be fed as input to the computer's printer port. The comparator's reference voltage is set by means of a trimmer potentiometer allowing the precise adjustment of the sensor's triggering condition.

The sender-side software is written in the *Processing* programming language (Read and Fry 2007). The image to transfer (a human stick figure) is stored in a rectangular array of Boolean values, which is initialized from an image drawn in the source code using so-called ASCII art. This provides a visual representation of the figure in the code, as can be seen in the following code excerpt.

```
// Initialize the img array from its textual image    representation
String simg =  " # " +
              "###" +
              " # " +
              " # " +
              "# #" ;
for (int r = 0; r < rows; r++)
    for (int c = 0; c < cols; c++)
        img[r][c] = (simg.charAt(r * cols + c) ==    '#' );
```

Fig. 7.8 State transition diagram depicting the program's operation

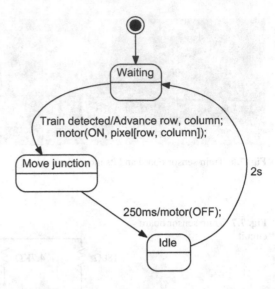

Programs written in *Processing* have by default an implicit event and drawing loop, which continuously calls the program's *draw* function. To guide the program's control flow I coded the operation of the *draw* function as a simple state machine (see Fig. 7.8). The program starts its operation in the *Waiting* state. Once the detector senses a passing train, it advances the row and column that must be sent, turns on the motor in the appropriate direction, and initializes a timer. In 250 ms the state becomes *Idle*, indicating a state where the program waits for the trim to pass through the function. Two seconds later, the program will enter the *Waiting* state, waiting for the train to make its next round.

The *draw* function will also redraw the image being sent, flashing the pixel in transit with a duty cycle of 500 ms. This is done with the following code.

```
for (int r = 0; r < rows; r++)
    for (int c = 0; c < cols; c++) {
        if (r == s endingRow &&   c == sendingCol &&
            ((millis() /   500 ) & 1) ==   0) {
            if (img[r][c])
                // Flash ON pixel
                drawFill(color(   180 , 64 , 64 ));
            else
                // Flash OFF pixel
                drawFill(color(   64 ));
        } else if (img[r][c])
            // Draw ON pixel
            drawFill(color(   255 , 0, 0));
        else
            // Draw OFF pixel
            drawFill(color(   0));
        drawPixel(r, c);
    }
```

Fig. 7.9 The receiving computer

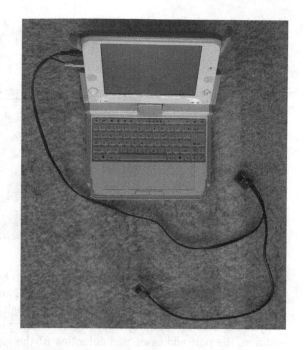

The *Processing* language is based on Java, which cannot directly access I/O ports. Complicating matters further, modern versions of the Windows operating system, do not allow any user-mode program to access I/O ports. These two restrictions were lifted by downloading and installing the *ParallelPort* Java class, which uses the Java Native Interface (JNI) to access the parallel port, and the *UserPort* device driver, which allows user-mode programs to access the I/O ports.

3   Receiver Implementation

The receiving computer is based on a late-prototype of the One Laptop per Child XO-1 computer, in order to demonstrate the platform's effectiveness as an experimentation and teaching aid (Fig. 7.9). Two sensors are used: a vertically mounted one to detect the approaching train and a horizontally mounted one to sense the value of the bit that the train is carrying on its top (Fig. 7.10).

In contrast to the circuit of the sending-end computer, the receiving-end sensor circuit (Fig. 7.11) does not convert the phototransistor's analog voltage level into a digital signal. Instead, it utilizes a hardware design feature of XO-1 that allows its audio input to be used as a sensor for analog values. This feature is there to aid experimentation and does indeed simplify the sensor's connection. Thus, the two sensors are directly connected to the XO-1 audio input.

The receiver's software was implemented in the Squeak (Ingalls et al 1997) Etoys environment (Gaelli et al 2006) as ported to the XO-1 laptop (Freudenberg

Fig. 7.10 The rotating device mounted on the train and the bit sensor

et al 2009). The XO-1 laptop port allows Etoys to be used from within the XO-1 shell by adopting the look and feel of other OLPC activities, by providing support for the persistence of programming projects through the environment's journal facility rather than files, and by allowing the sharing of projects between pupils. In addition, the port addresses particularities of the laptop's hardware, such as the higher screen resolution, a processor with relatively low performance (433 MHz), a color scheme that is not based on sub-pixel color components, and the ability of the audio jack to be used for sensor input. This last feature requires the analog to digital converter hardware to be switched from AC mode into a DC mode thereby removing the filtering of an audio signal's DC component. A new object, called "World Stethoscope" supports the use of the microphone jack as a sensor.

Given that the Etoys' World Stethoscope object provides only a single input it is used to handle the input both from the sensor detecting the train's approach and the sensor detecting the bit's value. This is done by implementing in software what is effectively a frequency modulation (FM) decoder. The software waits for a fixed time interval after a train passes to see if a "1" bit is detected by the bit value sensor. If the sensor does not detect such a value, the software assumes that the value is "0".

The receiver is implemented as a state machine (see Fig. 7.12 left). In the wait-Train state the software waits for a train to approach. When the sensor's read out value rises above a pre-determined threshold, the software enters into the waitZero state, where the software waits for the train to move away from the train sensor. This happens when the sensor's read out value falls below the threshold value. At that point the software enters the waitBit state, where it zeroes a tick counter and the bit value and waits for a bit sensor value to appear. If the sensor's read out value rises above the threshold, the bit value is set to true (1). After 30 ticks, a dot (pixel) with the recorded bit value is added to the image reconstructed on the screen, and the image's current coordinates are updated. (The image is transmitted as a series of pixels; the receiver has hard-coded the image's dimensions.)

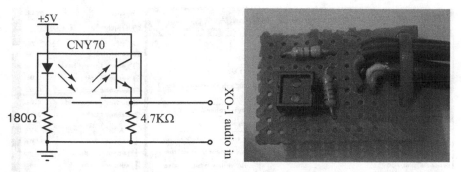

Fig. 7.11 Receiving-end sensor circuit and board

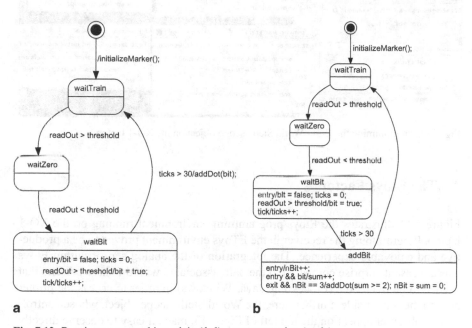

Fig. 7.12 Receiver state machine: plain (*left*), error correcting (*right*)

The experiment also allows the demonstration of error detection and correction. Under this scheme, each pixel is transmitted three times; at least two "1" values indicate a pixel with an "on" value, while at least two "0" values indicate a pixel with an "off" value. Errors can be easily introduced by manually manipulating the direction of the train's L-shaped data carrier. The corresponding state machine diagram is illustrated in Fig. 7.12 (right). An additional state, addBit, counts the number (nBit) and sum (sum) of the bits received. When three bits have been received (the train has passed three times from the receiving station), a pixel is added based on the bits' sum value, and the two counters are zeroed.

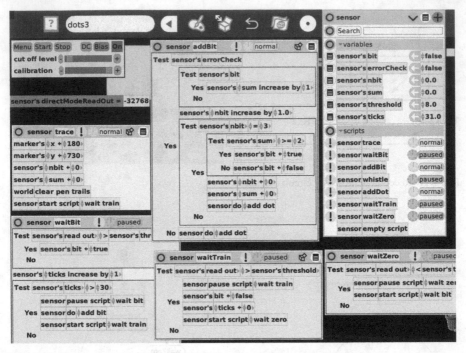

Fig. 7.13 Programming the Etoys World Stethoscope object on the XO-1 laptop

4 The Etoys Factor

Figure 7.13 illustrates the Etoys programming environment running on the XO-1 laptop. Programming the receiver in the EToys environment proved to be a productive and enjoyable experience. The integration of the analog sensor within Etoys was a pleasant surprise compared to the pain associated with the three barriers that needed to be overcome (Processing, Java, Windows) in order to access the printer port on the sender side. Furthermore, the World Stethoscope object, whose controller and observer appear on the top left of Fig. 7.13, made it easy to observe directly the sensor's value and create an appropriate threshold for detecting objects.

Interestingly, there is a one-to-one correspondence between each state of the state machine that describes the receiving software functioning and an Etoys script. This makes it easy to understand how the state machine functions and to observe its operation. Scripts in Etoys can be running or paused. Pausing a script and starting another one is the equivalent of moving from one state to another. During development it was easy to set the machine to a specific state, simply by clicking on the corresponding script's clock icon to have the script begin "ticking" and thereby made active.

Similarly, all the variables and states associated with the receiver were easily visible as a sensor pane (top right on Fig. 7.13). This made it easy to debug the software during development, but, more importantly, it also allowed me to explain its operation when the experiment was demonstrated to children. For instance, it was fascinating to see the tick count begin to decrease once the train passed through the first sensor. In addition, explaining the error correction functionality without seeing the corresponding variables would have been futile.

5   Experience and Lessons Learned

The experiment was exhibited in a scientific experiment contest organized by the Cooperative Association for Internet Data Analysis (CAIDA) non-profit organization and the Eugenides Foundation in Athens, Greece. There it was awarded one of the contest's three prizes. The stated goal of the experiment was to demonstrate to children the basic elements of information theory (Shannon 1948): conversion of an image from analog to digital form, the transmission of information as bits, the reconstruction of an image from digital bits, and reliable communication in the presence of errors. The experiment's Lego construction and the rapidly moving train attracted many children to the stand. A large proportion stayed to see the experiment in action, and many asked questions and appeared to understand the concepts behind it. Given the types of parents who take the children to science fairs, it was no surprise that many parents also asked questions, not only about the theory about behind the experiment, but (mostly) concerning its implementation.

The experiment's construction and operation proved to offer something for everyone. A toddler would look at the rotating train and listen it whistle as it passed from the receiving station, and even help the construction of the tracks. An older child could help with the other parts of the Lego construction, and observe how a train could transmit a picture pixel-by-pixel from one computer to the other. Those with an interest in programming could observe the receiving end's script operation, while the more mathematically inclined would appreciate the error correction algorithm. The design and soldering of the hardware as well, the sender implementation in the Processing language, and (probably) the design of the receiver software were tasks that required an adult.

A need to involve an adult in the process should not be taken as a negative verdict on Etoys. There are many interesting and worthwhile activities that children without the help of an adult can perform in Etoys in general, and using the World Stethoscope object in particular. Examples include games, animations, demonstrations, and setups to react to the outside world. For instance a child could connect a photoresistor to the XO-1's input and have a sun on a scene rise and set based on the light the resistor receives. Or it could build a simple alarm, by monitoring a switch to trigger a horn sound. Nevertheless, the World Stethoscope object could benefit from some polishing to make it more child-friendly. The elimination of terms like "Bias", "DC", and "calibration" from its user interface and operation, would go a long way toward this direction. Instead of the various adjustments, the corresponding operations could be performed transparently behind the scenes based on the input the object receives.

Etoys proved, once again (see Spinellis 2008), to be an interesting platform to demonstrate non-trivial programming concepts. Alan Kay, a driving force behind Squeak on which Etoys runs, has written that to learn science we "have to find ways to make the invisible visible" (Kay 2003). This credo was made possible at many levels during the development of the experiment. Most obviously children could see the bits carried around on the train making their way into the picture. At a deeper level, elements of the receiver program were also visible. The state machine's states, rather than being hidden behind an opaque variable, as is usually the case in such implementations (Thomas and Hunt 2002), were visible as separate scripts implementing the transitions. More impressively, the active state was also clearly visible during the operation as the currently running Etoys script as a "running" script, whereas the scripts associated with the other states appeared as "paused". Similarly visible were the values of all the program's variables (see Fig. 7.13). This allowed me to demonstrate to observers how the physical world (the train) interacted with the program: how the program's variables changed as the train passed through the sensors and the scripts run and paused.

The experiment's modular construction allowed each part to be designed, constructed, and tested in isolation, allowing the experiment to be gradually implemented in a period of months. The modular construction also made it easy to transport the experiment: all parts could fit in a suitcase and could be readily assembled on site. Switching the junction motor with relays proved to be an easy and reliable method, in contrast to solid-state approaches tried for another task. In early phases of the construction the clicking of the relays provided reassuring feedback that the wiring and the software were working correctly. In contrast, the detection of objects using infrared sensors proved a tricky affair. Various integrated and transmitter/receiver pair sensors were tried with mixed success. Even the sensor used proved to have difficulty operating within the relatively large distance tolerances of the Lego train set. (The sensor's intended use was optoelectronic encoder assemblies, such as index and coded disk scanning.) In the end, an adequate signal level was obtained by covering the Lego parts that the sensor should detect with aluminum foil.

Given the experiment's vertical integration and its diverse elements (large scale, normal scale and Technic Lego parts, electronic sensors and actuators, laptops, software platforms, software code) the instances where design choices obviated the need for additional work or elements made a considerable difference to the project's viability. These choices included the saving of two external power supplies by obtaining power from each laptop's USB port, the ability of the XO-1 to use its microphone input as a DC voltage analog to digital converter, and the integration of that converter within the Etoys platform. Although each saving may appear trivial, all together they can add up contributing to a project's death by a thousand cuts. As a counterexample, accessing the printer port under Processing involved code written in C and in x86 assembly language.

Attendees asked about other projects that could be implemented in a similar way, and provided some interesting ideas. For instance, one suggested that the receiver could be implemented using a single sensor that would first detect the train and then the presence of a "1" bit. In general, mapping microscopic and extremely fast phenomena onto an observable experiment can have many other applications. Here are some examples. Demonstrate the functioning of hard-disk secondary memory by physically storing bits (e.g. Lego blocks) on a large rotating disk. Make a concrete application out of the hard disk memory by storing a game's high score. Show analog-to-digital and digital-to-analog conversion by loading a discrete number of blocks into a carriage. Explain printing by taking pictures with a physical camera and printing them using a motor-driven carriage driving a drawing servomotor. The possibilities are endless.

Perhaps the most important lesson I learned from the experiment was the importance of accessibility. The experiment provided affordances that allowed direct manipulation of many aspects of its operation. Children could readily observe the signal carried atop the train, stop the train to delay the signal, and change the value of the carried bit to see the effect on the receiving and or on the error-correction system. On the XO-1 Etoys end children could also see the software's operation laid bare like a cross-cut of a working car engine. Although I started with this as a design goal, I feel fortunate that this aspect worked better than my most optimistic expectations.

References

B. J. Allen-Conn and Kim Rose. Powerful Ideas in the Classroom Using Squeak to Enhance Math and Science Learning. Viewpoints Research Institute, 2003.

Claude E. Shannon. A Mathematical Theory of Communication. Bell System Technical Journal, 27, pages 379–423 and 623–656, July and October, 1948.

Dave Thomas and Andy Hunt. State Machines. IEEE Software 19(6): 10–12. November/December 2002.

Diomidis Spinellis. The Antikythera mechanism: A computer science perspective. IEEE Computer, 41(5):22–27, May 2008. (doi:10.1109/MC.2008.166)

Fairchild Semiconductor. LM339/LM339A, LM239A, LM2901: Quad Comparator. Revision 1.0.5, 2012. Available online www.fairchildsemi.com/ds/LM/LM2901.pdf.

B. Freudenberg, Y. Ohshima, and S. Wallace. Etoys for One Laptop Per Child. In C5 '09: The Seventh International Conference on Creating, Connecting and Collaborating through Computing, pages 57–64, 2009. (doi:10.1109/C5.2009.9)

M. Gaelli, O. Nierstrasz, and S. Stinckwich. Idioms for composing games with EToys. In C5 '06: The Fourth International Conference on Creating, Connecting and Collaborating through Computing, pages 222–231, 2006. (doi:10.1109/C5.2006.20)

Dan Ingalls, Ted Kaehler, John Maloney, Scott Wallace, and Alan Kay. Back to the future: the story of Squeak, a practical Smalltalk written in itself. In OOPSLA '97: Proceedings of the 12th ACM SIGPLAN Conference on Object-Oriented Programming, Systems, Languages, and Applications, pages 318–326, New York, NY, USA, 1997. ACM Press. (doi:10.1145/263698.263754)

Alan Kay. Our Human Condition "From Space". In (Allen-Conn and Rose, 2003) pp. 73–79.

Newton Lee. Interview with Nicholas Negroponte. Computers in Entertainment, 4(1):3, 2006. (doi:10.1145/1111293.1111298)

Casey Read and Ben Fry. Processing: A Programming Handbook for Visual Designers and Artists. MIT Press, Cambridge, MA, 2007.

Texas Instruments. ULN2002A, ULN2003A, ULN2003AI, ULN2004A, ULQ2003A, ULQ2004A: High-voltage high-current Darlington transistor arrays. December 1976, revised March 2012. Available online www.ti.com/lit/ds/symlink/uln2003a.pdf.

Vishay Intertechnology. CNY70: Reflective Optical Sensor with Transistor Output. Document number 83751. Revision 1.8, July 2012. Available online www.vishay.com/docs/83751/cny70.pdf.

Chapter 8
The QUARTIC Process Model for Developing Serious Games: 'Green My Place' Case Study

Benjamin Cowley

1 Introduction

The potential of "serious games" as tools for learning is recognised as an exciting possibility for Technology Enhanced Learning (TEL), but as a comparatively new domain for software development the possible benefits are often blocked by serious barriers, including uncertainty in how best to specify simulations, and a lack of repeatability even with successful products. Part of the problem lies in finding ways to successfully marry game design with pedagogical theory (or, equivalently, to integrate pedagogy design with a game's core mechanics). Attempts to deal with these issues must face the daunting complexity of the systems involved, and the relative incompatibility of the methods used in each domain. Game design is itself far from an exact discipline, and one that usually approaches its problems from the space of entertainment software, leading to a focus on player satisfaction rather than effective education. Finding systematic solutions to the problems of educational game design may serve to make the associated problems more tractable.

In principal, it should be possible to align pedagogical and game design aspects, since both are focused on the same domain—player/learner interaction with the educational game. When learning through playing such games, the learner (in the ideal case) undergoes an engaging experience that contributes to the development of her competences (Kolb 1984). However, in order to achieve this in practice, a serious game must be designed, from the ground up, to harmonise the entertainment elements and the educational elements around the specifics of the pedagogy involved. These two goals inevitably conflict in so much as games for entertainment tend to succeed for reasons unrelated to their potential for education. For instance, *Bejeweled* (PopCap 2001) entertains a great many players who learn how to spot

B. Cowley (✉)
Cognitive Brain Research Unit, Cognitive Science, Institute
of Behavioural Sciences, University of Helsinki, Helsinki, Finland

Brain and Work Research Centre, Finnish Institute of Occupational Health,
Helsinki, Finland
e-mail: ben.cowley@helsinki.fi

N. Lee (ed.), *Digital Da Vinci*, DOI 10.1007/978-1-4939-0965-0_8,
© Springer Science+Business Media New York 2014

transposable elements in a grid to make matching patterns, but this skill has limited or non-existent application outside of the context of match 3 games.

Another challenge for serious games that is not often addressed is that the prevailing trend is for literal simulation, which often results in games with a high level of complexity and a correspondingly narrow audience. Complex simulations are important serious games applications, but not necessarily the optimal choice for educational software. Commercially successful entertainment simulations such as *The Sims* (Maxis 2000) and *FarmVille* (Zynga 2009) invariably streamline the simulation aspects of their gameplay in order to make them accessible, and if educationally-motivated serious games wish to appeal to a similarly wide audience they too must reflect the player satisfaction goal of an easy-to-learn system. However, such simplified systems no longer fully reflect the situation represented—their educational value is curtailed.

Furthermore, recognising that a simpler system is a requirement for accessibility is often difficult for the programmers and engineers who build the games—they tend to have high tolerance for and enjoyment of complexity, and detailed simulation games. This creates a pre-existing bias towards this design approach that might not reflect best practices. There are also inherent difficulties involved in conceiving of ways to disseminate serious topics through an apparently frivolous medium. Both of these issues relate to cognitive bias of the designer or developers, and therefore need to be dealt with in the process model, not by relying on open-mindedness to win out. Both these issues are discussed in more detail in Cowley et al. (2011).

This paper presents a process model for developing a client-focused educational game that relates the pedagogical processes, being in this case design for awareness and behaviour change training, to the game design process. These two distinct but interrelated processes are connected by a recursive dependency that works to keep the needs of each process integrated and balanced. Within this process model, software development is both guided and repeatable, with specified requirements and pedagogical goals propagated throughout the resulting design process. This model has been applied in the development of the serious game *Green My Place*, and this game is explored through the lens of the model throughout the paper.

1.1 Green My Place

The award-winning[1] serious game *Green My Place*[2] is aimed at achieving behaviour transformation in energy awareness. It takes the form of a massively multiplayer online (MMO) game played within a web browser, focused on five different building locations around Europe which are instantiated as teams in the game.

---

[1] First 'European Best Learning Game Competition', first place prize in the category 'Best Non-professional functional game'.

[2] To Access Green My Place, go to http://greenmyplace.net.

Teams compete over one year to become the most energy efficient. Score in this competition is comprised of two elements:

- Real-world energy sensors that measure energy use in each building.
- Additional points accumulated by players join teams that correspond to each of the buildings

Green My Place is the keystone in the user interface of the SAVE ENERGY European Project (CIP-ICT-PSP-238882 PROJECT)[3]. SAVE ENERGY addresses the challenge of behaviour transformation through the use of Information and Communications Technology (ICT) in order to encourage superior energy efficiency in five public buildings in five European cities—Helsinki, Leiden, Lisbon, Luleå and Manchester.

Participants in *Green My Place* play simple, fast eco-action based mini-games, take quizzes and learn about energy efficiency from sources on the web. By participating in the games and additional activities, players contribute to the winning of Awards that improve and upgrade the virtual representation of their team's building. The acquisition of Awards in the MMO provides a bonus towards the main score (which, as noted, is based upon the energy savings of their team in the relevant building). Thus, five teams, each corresponding to a pilot building, compete for one year to "green their place". This design also seeks to address one of the drawbacks of the serious game concept: where the game is at once too compelling and yet not tightly integrated enough to pedagogical goals, so that learning devolves totally to the mechanics of the game and not any transferable skills or behaviours.

This paper focuses on the process used to create a game specifically intended to promote behaviour change, which is a significantly different scenario to games which teach competences—this topic was dealt with in Cowley et al. (2012). One may still speak of a "lesson to be learned"; however the object in this case is not to impart skills but to raise awareness, offer solutions and elicit motivation for the desired behaviour change. Whether this is judged a limitation of the *process* cannot be assessed from this one instance—it is possible that the process is equally applicable to software development for other learning outcomes, and there is no apparent reason that it could not be applied in this way. However, because discussion in this paper explicitly concerns *Green My Place* the background of behavioural change should be borne in mind.

1.2   QUARTIC Process Model

The QUality Assuring Recursive TEL Instruction co-Creation (QUARTIC) process model shown in Figure 8.1 is a recursive spiral model of game design, incorporating best practice for behaviour-change pedagogy development.

---

[3] For more on SAVE ENERGY, see http://www.ict4saveenergy.eu.

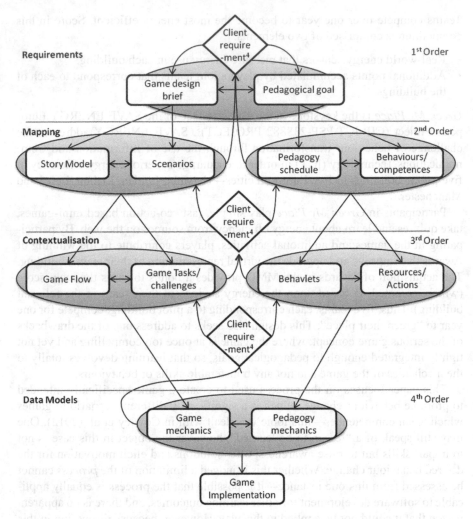

Fig. 8.1 QUARTIC. Our process model for developing serious games, specified for Green My Place. (Client's original requirement provides the schematic of what must be produced. The designers'/developers' expertise and judgement must fill out the details, but always stepping back up a level to check that progress is in line with requirements)

The central concept is that a non-prescriptive TEL game experience can be built using an iterative cross-correlation between *goals* (lesson to be learned/behaviour to be changed) and *means* (scenarios and game mechanics for teaching). Starting from a given client[4] requirement—formulated as either behavioural goal(s) or playable scenario—the design team develops one of the parallel streams, the *story stream*

---

[4] We define 'client' as the commissioners of production of the serious game – be it a company, institution or end-user.

(the left-hand stream in Figure 8.1, focussed on game development) or the *behaviour stream* (the right hand stream in Figure 8.1, focussed on behaviour modelling), which interrelate to produce the final software. While this case study focuses on behaviour change in the energy domain, the same principles could be applied for competence assessment, as discussed elsewhere (Cowley et al. 2012).

The practical streams of development begin with client requirements (diamond shapes), and are composed of sequential interlocking (oval shapes) design tasks (rounded-cornered rectangles), which nevertheless are part of a recursive structure (i.e. each stage affects the parallel stage in the other stream). Consequently, as each stage is completed the process descends one stage before returning to the previous stage of the parallel stream in order to perform requirements analysis[5] and develop the next stage for that stream. With given methods at each stage, developers can step through several orders of complexity with confidence and clarity, and not become "lost in the data".

The QUARTIC model is broken into four orders of complexity, marked in the figure by horizontal lines, which define the order of working and correspond to stages in the development as follows:

1. **Requirements (1st Order)**. The key picture of client requirements will provide either a *behavioural goal*, or a *scenario fragment*, from which the next stage can be derived. The next stage provides a stable basis from which to go back and generate the key picture of the first stage in the other stream. Thus the client only needs to provide a single data, or data set—either the rough pedagogy description, or the rough scenario outline.
2. **Mapping (2nd Order)**. To obtain a complete but non-contextualised schedule of the pedagogy to be taught, a map of behaviour must be built. Similarly the key elements of the story, to be used for the educational game scenario, are outlined by use of a story model.
3. **Contextualisation (3rd Order)**. This stage is needed to tackle the context of the behaviour and scenes. Specific models of potential energy saving behaviours, known as Behavlets are defined. For instance, a Behavlet dealing with "vampire power" from phone chargers specifies a potential 25–30 % energy saving from unplugging phone chargers when not in use.
4. **Data models (4th Order)**. In the fourth order/final stage, the previous stages influence the instantiation of entities within the scenes which may affect the learner's experience—data, social information, in-game tasks or game mechanics. This stage is also when plans for further prototype iterations can be solidified based on testing.

The use of the model is relatively simple, governed by three principles. First, follow the arrows one at a time; second, observe the *original* client requirements as checkpoints; third, complete all the work in both streams of order x before completing both streams of order $x+1$ (in other words never proceed to order

---

[5] The client's original requirement provides the schematic of what must be produced. The developers' expertise and judgment must fill out the details, but always stepping back up a level to check that progress is in line with requirements.

$x+2$ without completing both streams of order x). Potentially the most interesting and difficult aspect of the model is in the interaction between the left *story stream* and the right *behaviour stream*; how should the developer unify these disparate activities? In this, QUARTIC is meant to guide at a high level. By observing the principles just given, the developer is forced to have a thorough understanding of their own status and direction in each stream, *before* going too far in developing the other stream. For example, as Sect. 3 shows we began developing *Green My Place* with a pedagogical goal. Before proceeding to develop the specific tools that would facilitate our pedagogical schedule (Behavlets), we were forced to come up with at least a game design brief; to know what structure of entertainment would 'wrap' the serious part.

Observing the model with these principles in mind, the purpose of the oval containers stands out—they represent *tightly-coupled* development areas. In order to move from one stream to another (i.e. to develop in *parallel*) or back up one stage (i.e. to *revise* work), one must normally pass through a requirements check, to review what has just been produced. Exceptions to this are within ovals, where design tasks are so related that it is possible to apprehend the outcome of design decisions.

The next section describes the developmental background and some of the state of the art in respect to this process model and *Green My Place*. The middle sections step through the orders of the QUARTIC model, using the *Green My Place* case to illustrate how each stream at each order is developed and integrated with its relevant context.

Within the section for each order are subsections discussing the two streams (story and behaviour)—in this case, the behaviour stream is always treated first because the development project began with requirements for behaviour change. Normally, where requirements favour one stream over the other, it would be sensible to deal with the favoured stream first, at each order, as this helps keep the design process orderly. Finally Sect. 7 presents an evaluation and in Sect. 8, conclusions are presented and potential for future work discussed.

2   Background and State of the Art

The QUARTIC model was originally developed as a solution to the particular design requirements in two European projects on serious games, SAVE ENERGY and TARGET (IST 231717). As this institution was involved in developing elements of both games, it became apparent that a clear process for integrating partners' disparate expertise in pedagogy and game design was lacking. Excessive time and international travel was required just to arrive at a common understanding of the development plan. Many bottlenecks were met while chains of partners waited on each other for input. While facilitating progress past these obstacles, it became clear that a new kind of software design process model was needed, one which would give ready structure while allowing sufficient freedom to innovate. As mentioned, this model derives from the peculiar nature of the serious games development domain and as such has been requirements-driven.

However it has also been informed by existing process models, such as the Cross Industry Standard Process (CRISP) model for data mining (Squier 2001) or the evidence-centred design approach to construct educational assessment procedures (Mislevy et al. 2006), or even the spiral life cycle model (Boehm 1995). In a manner similar to CRISP, the QUARTIC process is intended to help encourage a reliable and repeatable development environment, to give a uniform framework for guidelines and for documenting experience, while at the same time being flexible enough to account for different needs. As with the CRISP model, QUARTIC is not a one-shot process but reverberative between stages, and thus implies a working model that is agile and communicative, such as Scrum (Schwaber and Beedle 2002). Indeed Scrum practices were used as part of the Green My Place development.

As with CRISP, QUARTIC is intended to be

- Non-proprietary
- Application/Industry neutral
- Tool neutral
- Focused on client issues
- A framework for guidance
- Provide an experience base of prior cases/templates for analysis

A necessary requirement for QUARTIC to work as a development guide is a description (for each stage in both streams) of the implementation steps and a description of the relational schemas in detail. This provides an implementer a framework for moving down an order by progressively developing a general model into a more specific one. An important part of this process is knowing when and how to 'recurse', that is, to move back up an order and across to the parallel stream in order to seed the generation of *that* content with the output already achieved in the other stream. This recursion works within a larger process of recursion, where the all four orders are visited within a single loop.

To visualise this larger process, it may help to map the QUARTIC model to the spiral design model, shown below in Figure 8.2 The spiral design methodology is a risk-managing iterative approach, meaning the project is conceived, designed, prototyped, tested, and the test results are put into the next iteration of the same steps. The spiral model is thus an evolutionary process. The design process starts in the centre of the spiral with an idea. It proceeds outward clockwise through each of the four phases of design. Every return of the spiral to the "define" position starts a new iteration. Iterations produce a deliverable or prototype, and each prototype is closer and closer to the final production model in complexity and degree of completion.

In our mapping, the process begins with an idea for a game or pedagogue, and continues as follows:

1. **Determine Objectives**—in the QUARTIC process, the Concept of Requirements is developed into a requirements plan within the first order. When the first order is visited again, the developed product can be checked against requirements again, which themselves can be evolved at each iteration.

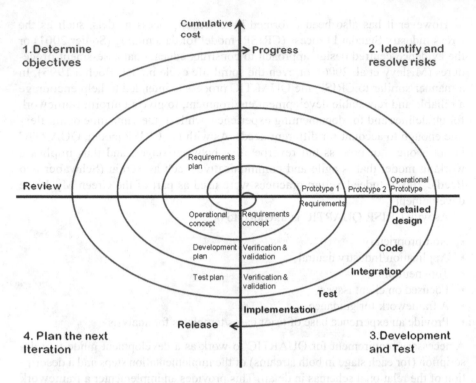

Fig. 8.2 The spiral model of design. This consists of four phases—1. Determine objectives = QUARTIC order 1;2. Identify and resolve risks = QUARTIC order 2;3. Development and test = QUARTIC order 3 and 4. Order 4 also overlaps somewhat with spiral phase 4, Plan the next iteration. Some aspects of the spiral model are implicit in the QUARTIC model, but not explicit (as it is not the primary aim of QUARTIC to map this model). Image adapted from Boehm (1995)

2. **Identify and Resolve Risks**—design starts here by writing the rules for the core mechanic and making the basics workable. The QUARTIC second order allows this without yet incurring the added overhead of contextualisation.

3. **Development and Test**—build a working model and start to get the look and feel of the physical features of the game. Contextualisation occurs in the third QUARTIC order, by matching game models to the domain of the serious content to be taught. QUARTIC order 4 overlaps slightly between this phase's implementation, and the next phase.

4. **Plan the next iteration**—in QUARTIC, order 4 also involves understanding what may need to be changed in the existing product.

Under the umbrella of educational games are two significant areas that must be harmonized—design/development of the technology, and learning support for the subject matter. The most relevant literature is that which addresses the *harmonisation* rather than the separate areas. In previous similar efforts Raybourn (2007) presents a design method for building adaptive training systems, based on prior work building simulations. Also Marsh et al. (2006) illustrate their serious game design approach using the "hierarchical activity-based scenario" (HABS) to provide a theoretical framework and a data mining/metrics approach for continuous evaluation of the product. Such existing methods as we have found are, we feel, orthogonal to the proposed model, being less focused on converging disparate design areas, and more on pursuing specific design agendas such as creating simulators. This implies that QUARTIC could plausibly work alongside such prior approaches. The following sections cite more work on educational games, behavioural change and contextual learning.

2.1 Educational Games

Educational games have received mixed support from academic studies (Egenfeldt-Nielsen 2006; Gee 2006; Wong et al. 2007)—for some time it was debated whether they worked or not—but as more research was reported on play in general (to show that some form of learning is almost *always* involved (Koster 2005)), it has become clear that the better question is whether a serious game is *on target*. Does a particular game teach retained, transferable skills which are the ones intended by the designers?

Squire argues that educators and curriculum designers should pay attention to video games as they offer "*designed experiences*, in which participants learn through a grammar of *doing* and *being*" (Squire 2006). He advocates that educators designing games should shift the focus from *delivering content* to *designing experiences*. Educational games support a situated context for learning in a virtual world because when you learn by playing a game, you apply that learning immediately in the game and move on to learning new skills (Gee 2003). Game scenarios and characters in the game that reflect the real world will enable a near-transfer of knowledge.

The learning principles that naturally occur in games are covered comprehensively by Gee (2003). (Cowley et al. 2011) give a clear example of the application of such principles to the design of *Green My Place*. Fundamentally, games facilitate learning (Gee 2003), but the lesson learned may be well hidden even to the game designer (Cook 2006; Schell 2008), if they are not explicitly aware of both the design of the play experience and also the pedagogy. McGinnis et al. (2008) draw a parallel between the factors describing student engagement and those involved in game play, which illustrates the elements necessary for creating game-like engagement in pedagogy.

People differ in their learning, social and play styles (Goldberg 1993). For each person, some forms of learning are more entertaining than others (Bateman and Boon 2005), and this represents a rich source of variation which an adaptive game may monitor to provide relatively optimal solutions (Charles und Black 2004; Cowley et al. 2006). Integrating scientifically-informed analysis with educational games design can therefore deliver software which meets the mutual goals of entertainment and education.

2.2 Behaviour Change

In terms of efficiency of energy-use behaviour, intervention studies have been carried out in many parts of the world. Abrahamse et al. (2005) and Darby (2006) have reviewed a substantial number of interventions, while Ehrhardt-Martinez (2008) extracts the policy lessons from a substantial literature review. Notable single studies for the purposes of SAVE ENERGY include (Abrahamse et al. 2007; Lockwood and Platt 2009; Petersen et al. 2007).

Behavlets

A new concept developed for this project is the Behavlet (Moutinho et al. 2010), which represents a critical incident of energy-use behaviour that can influence a situation toward greater energy efficiency. The idea behind the Behavlet is to represent the simplest behavioural change possible in a given context, and to also represent the consequences of that change (and any additional information pertinent to that behaviour). The key elements of each Behavlet are the *Resources*, which specify what is involved in the behaviour, the *Action*, which specifies what is to be done with the relevant resources, and the *Outcome*, which specifies the potential benefits of the relevant behaviour.

A simple example can be found in the case of "vampire power" when a phone charger is left plugged into a socket. Even if the phone is not connected to the charger, the charge continues to draw power, generating considerable wasted energy. The Behavlet in this case specifies mobile chargers as the resources, unplugging when not in use as the action, and 25–30 % energy savings as the outcome.

Because each Behavlet represents an isolated behavioural activity with the potential for the desired benefit (in this case, saving energy) they represent "atoms" of behavioural change that can be used to construct both games and scenarios for superior energy efficiency. Crucially, Behavlets can be researched on the basis of empirical data and then used as an input for the story stream, producing game designs that relate to actual situations *without* the necessity of detailed simulation. In the case of the "vampire power " Behavlet, a game that puts the idea of unplugging phone chargers that are not in use into the mind of its player can (if this behaviour change is successfully attained) achieve the goal of energy saving.

2.3 Contextual Learning

One of the requirements for the success of game-based learning has been identified as linking the game to the learning or curriculum objectives (Carpenter and Windsor 2011). Several researchers advocate that learning must involve more than the transmission of knowledge, but must instead encourage and happen through rich contexts that reflect real life learning processes (Lave and Wenger 1991).

The contextual learning pedagogy has been developed to respond to these increasing demands posed by the rapidly changing and complex operational environment. Karweit (1993) defines contextual learning as learning that is designed so that students can carry out activities and solve problems in a way that reflects the nature of such tasks in the real world. This draws ideas from John Dewey, Kurt Lewin, David Kolb and other work on experiential learning and the research on the effectiveness of learning in meaningful contexts.

3   Requirements (1st order)

The QUARTIC process begins with a client requirement. The game may need to teach a particular behaviour, or it may need to relate to a particular scenario, in which the exact behaviours of interest have not been defined by the client but are implicit. In either case, the process is the same:

1. **Develop Requirements in Primary Steam** Systematically develop the requirement given for one stream until a clear map is obtained for the 2nd order (mapping).
2. **Produce Rough Requirement for Secondary Stream** Use the map produced in (1) to derive the rough description of the requirement for the *other* stream, in consultation with the client when necessary.

For example, in the case of the *Green My Place* client (the SAVE ENERGY project), the main objective is to make use of ICT to transform the behaviour of users of public buildings regarding energy efficiency, through serious games and real-time information from sensors and actuators. SAVE ENERGY builds upon the Living Labs methodology (Oliveira et al. 2006) to provide an engaging virtual environment for users, citizens and policy makers to gain awareness, understanding and experience associated with energy saving attitudes. In this context, the problem of serious game design began with a prior requirement mainly relating to the *behaviour* stream, and thus this was the "Primary Stream" in the above description, while the *story* stream served as the "Secondary Stream".

Other contexts are equally possible, for instance, a competence map might be created first and then the story can be developed directly. It is also possible that *both* stream requirements might be obtained simultaneously, requiring them to be related to each other.

3.1   Pedagogical Goal

In order to achieve the project objectives, the game had to meet the following requirements and constraints.

- To educate people in ways that energy can be saved, via simple actions that anyone can conduct.

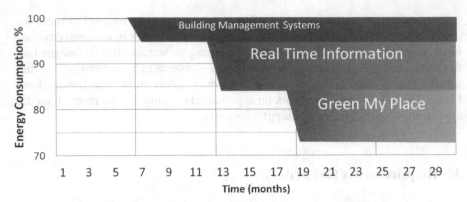

Fig. 8.3 Cumulative energy saving targets within the SAVE ENERGY project. The graph illustrates the sequential cumulative impact of the 3 active SAVE ENERGY methods, plotting energy consumption along the time axis. This figure also points to the evaluation strategy for each separate intervention of the project—for more see Sect. 7

- To increase interest in the pilot scheme buildings.
- To be an inherently enjoyable game for a sufficiently diverse audience

We can expand this into more specific terms:

- Targeting ≥25 % reduction in energy use attributable to SAVE ENERGY—the project goals are shown below in Figure 8.3.
- Achieve targets with 1 year of exposure of audience to game
- Utilise real-time energy monitoring data from the pilot buildings
- Integrate the game with real-time energy management (EM) and information-display (RTI) systems
- Deploy the same system across 5 different pre-defined audiences (both adult and young people) in 5 pilot buildings, in 5 different cities across Europe with 5 different languages

Since the goal of the SAVE ENERGY project is both education and behavioural change, it was important that the pilot scheme (which is the centre point of the endeavour) be rendered interesting to a wide audience. By itself, the pilot scheme is a worthy endeavour, but unlikely to attract the notice of the wider populace, thus risking a failure to capitalise on its achievements.

By re-framing the project in the context of a meta-game 'wrapper'—one that can be entirely external to the pilot scheme proper, and serves simply as a 'circus' to draw attention to what is being done, we aimed to better meet these educational goals. Furthermore, the meta-game created a motivating feedback loop between the pilot building workers and their community—actions taken in each of the pilot buildings had a meta-game effect which is meaningful to a large community of people. This helps strengthen the desire of the pilot scheme participants to work on saving energy. Consideration of the pedagogical goals allowed us to create the following simple mapping as a guideline for development:

- Activity Structure → Motivates
  - Content → Teaches
    - Targets → Enable
      - Rewards → Reinforce

From the perspective of this structured definition of our requirements (using the 'follow-the-arrows' principle in QUARTIC), we could then look at both the next order in the behaviour stream and the same order in the story stream *at the same time*. The outcome is a game design brief which more closely integrated with the pedagogical approach.

3.2 Game Design Brief

The brief for the SAVE ENERGY project was to create a fun and engaging game experience, while shaping the player's behaviour towards energy saving issues. The main focus of game play is to maintain awareness for the player of both energy efficiency issues in general, and the level of energy consumption in their local pilot building in particular.

Thus the game design focussed upon behavioural change that could be achieved by maintaining the presence of the game in the player's consciousness for as long as possible, without impeding their daily routine. Consequently, the aim was a *slow burn* experience, using the model of a massively social online game, or MSO (Lazzaro 2009). This takes the form of short duration educational "energy challenge" mini-games embedded in the ongoing social "Euro Team" meta-game. The game has a target audience (in priority order):

- Pilot building users: citizen, public servant, policy maker
- Non-users directly linked to pilot, e.g. school parents, families
- General public in pilot building local area

Players interact with the game over a year's duration—in order to obtain that much engaged game-play, the interaction needs to be self-perpetuating, light-weight and autonomously produce interest. In other words, the game must be greater than the sessions of play, since the player cannot be expected to invest very much into the each session.

In demographic terms, this audience is composed of arbitrary mass market (or "Casual") players i.e. the widest conceivable audience. In terms of meeting educational goals, this audience is preferred to a more narrowly constrained video game audience, such as would respond to a simulation experience (Cowley et al. 2011). No aspect of the game experience of *Green My Place* changes as a result of the nature of the player; all players are treated equivalently.

Given this structuring of the requirements step (the first order), we can describe the content required for the mapping step (second order)—bearing in mind that mapping in the primary stream (behaviour, in the case of *Green My Place*) should already have been at least cursorily examined while developing the secondary stream requirements (for *Green My Place*, story). This parallel attention to the two streams is a hallmark of the QUARTIC process, which aims to integrate the two individual streams as closely as possible.

4 Mapping (2nd order)

> When we notice similarities between two different systems … the comparison often begins
> in a literary manner. There is the simile, the more direct metaphor, conceptual models …
> The process continues, and begins to have the marks of a scientific method, when we try to
> develop rigorous formulations of the two conceptual models. (Espejo and Harnden 1989)

Mapping implies the translation from one means of description to another. The requirements step produces (usually) broad strokes descriptions in natural language, and this is insufficiently precise for the system to be built. As a result, the mapping step works towards a high detail formal description of processes and/or necessary components of the system to be built.

4.1 Pedagogy Schedule

The basis of the pedagogy in *Green My Place* is the interaction loop, which describes how players are to experience the material which teach them about energy saving. The loop is composed of multiple interrelated interaction *modes*, each of which the player enters voluntarily and which points to the next. The different interaction modes are:

- **Play** presents action/puzzle *mini-games* that relate to energy-efficiency, and which lead the player to learning materials on termination.
- **Learn** offers topical lesson-like materials, to provide supplemental information

These two modes exist embedded within the third mode: a *meta-game* which bridges both interaction types and induces a year-long *competitive/collaborative MSO* as a third, high-level type. Figure 8.4 shows an example of the interaction loop for *Green My Place*.

Each of these modes is described in terms of its pedagogical impact in the next three sections.

Meta-Game

The meta-game creates a motivating feedback loop between the pilot building workers and their community—the actions taken in each of the pilot building has a meta-game effect which is meaningful to a large community of people. This helps strengthen the desire of the pilot scheme participants to work on saving energy.

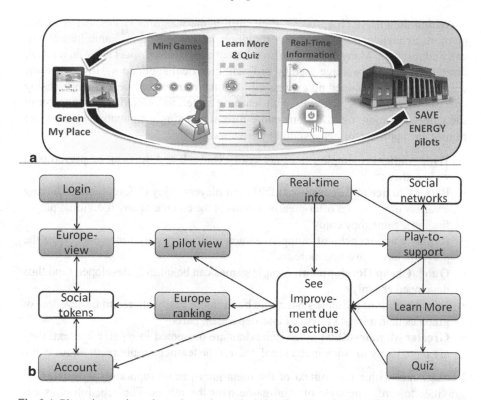

Fig. 8.4 Player interaction schematic. *A.* shows the hierarchy of modes, from bottom to top: meta-game play, learn more (leading to real-time information and thus the pilot environments). *B.* shows the flow of player interaction from first logon to meta-game, play, learning and back again

A comparison can be made to 'reality television', which did not reach its full potential (in terms of viewing figures) before discovering that combining game-like elements with the 'fly on the wall' format was a way to draw in a wide audience. In a similar way, the meta-game provides a structuring timeline which orders the content of the project in an easy-to-comprehend game-like fashion. This makes it easier for everyone participating in the project (both inside and outside of the pilot scheme buildings) to maintain focus on the project as it proceeds.

Play

The play aspect of the *Green My Place* motivating feedback loop is intrinsically self-motivating because—unlike many other serious game projects—the mini-games were constructed with entertainment concerns as primary and educational concerns secondary, with new mini-games were introduced on a weekly basis to further maintain interest in the meta-game.

Within the QUARTIC process, each mini-game was *always* connected to one or more Behavlets, but it was not a *requirement* that each mini-game be intrinsically educational. Many did contain an inherent educational aspect e.g. *Electrickery* directly teaches the wattage used by different electrical appliances. However some e.g. *AirconTroll* contained no express educational element, while *still* emphasising specific energy saving contexts, with learning materials offered on conclusion of the game. The more usual serious game format is a single simulation system; the mini-game approach offers several key advantages:

- **Wide Audience:** simple games are easily learned, and thus can be played by a wide audience
- **Better Chance of Engagement:** Different players enjoy different things; having a variety of games on offer greatly increases the chance of any individual player finding a game they enjoy
- **Replay Encouragement:** simpler games with short play times encourage the player to have "just one more try".
- **Quick/Cheap Development:** Simple games can be quickly developed (and thus developed cheaply)
- **Easy Variation:** a simple system can be far more easily altered to fit new data or graphics than a large system of interdependent parts.
- **Greater Memorability:** since Behavlets are presented in a game context, they are more likely to 'stick in the mind' then if the learner simply reads about them.

Furthermore, within the context of the meta-game, most supporters will naturally gravitate towards one style of mini-game over the others. The schedule thus engages different supporters each week, keeping involvement in the meta-game fluid and interesting.

Learn More

As noted, the mini-games may lack a key instructional component—the repetition of task-based interaction with material to be learned, something like the 'probing principle' of Gee (2003). Mini-games need to be short and novel which makes it difficult to repeat an instructional element often enough; while some designs afford this possibility, others do not.

The interaction loop uses the Learn More mode to solve this problem. The key concept of the learn mode is that the context is relatable to the recent experience of other modes—thus where the mini-game has a theme, the Learn More page should continue this theme, and extend it out the wider internet e.g. playing a mini-game about air conditioning leads to learning materials about air conditioning. This means that the type of content that is dealt with in frivolous play by the mini-game has a serious counterpart, which is addressed in part by key knowledge items, and in full by resources on the internet which can be linked to.

Table 8.1 Ten categories of energy saving content. At least one category is exemplified in each mini-game and thus represented in the Learn More page available after playing the game

1	Lighting
2	Air conditioning
3	Windows
4	Insulation
5	Behaviour
6	Appliance use
7	Energy efficient equipment
8	Automated environment controls
9	Heating
10	Movement

Green My Place mini-games are categorised by their energy-saving thematic content, which include the ten categories in Table 8.1. Based on this categorisation, after a game the player is given the option to proceed to the Learn More page, which is dynamically populated with Knowledge Items and Tips relating to the relevant category (as well as real time data from the team building).

4.2   Story Model

The story stream, tasked with providing the more frivolous side of the design, was first expanded from the design sketch by *literally* writing a narrative. The final iteration of this (as shown to players) is as follows:

- **Earth's in danger!** All over the World, people have the wasting sickness—they just use too much energy.
- **It can be better!** With Green My Place, 5 cities across Europe are improving—by learning to change their behaviour they can save real energy, not waste it.
- **Join us!** You can support 1 of these places by playing Green My Place—the serious game where Helsinki, Leiden, Lisbon, Luleå and Manchester compete to be the most energy efficient in 2011.
- **Let's Green My Place!**

Based on this narrative, the meta-game was planned to provide an on-going framework for the 'greening', with regular scenarios relating to key energy saving concepts provided within that frame. Some of the scenarios which were planned included:

- **Relative power (1)**—different appliances use different amounts of power. We expose this through the mini-game *Electrickery*, in order to raise basic awareness of the cost of various actions.
- **Relative Power (2)**—different settings use vastly different amounts of power. A commercial kitchen uses one or two orders of magnitude more power than a classroom, for instance. We planned to expose this using the comparable mini-games *Energy Dash* and *Kitchen Dash*.

- **Waste by neglect**—much of energy waste results from simply not setting the things we use to a lower power state once utilisation has finished. This is covered in the mini-games *WindowWatcher* and *Switch Search*.

5 Contextualisation (3rd order)

The third order, *contextualisation*, is where the 'serious' and the game must begin to converge tightly. This means that game interaction modes should be aligned to the context of the domain of interest, namely, the serious content which needs to be taught.

5.1 Behavlets

Desktop research and interviews were conducted by the SAVE ENERGY consortium to identify critical incidents in terms of energy efficiency behaviour concerning work places (e.g. unplugging any electrical device that's not being used). These critical incidents are basic actions that people could perform to affect directly or indirectly the building's energy consumption. Building upon these critical incidents, the concept of Behavlet was developed to represent a unit of behaviour transformation structured around actions, resources and outcomes—plus the most important contextual information.

The idea behind the Behavlet concept is to identify the means for making behaviour transformation possible in a context, through a loosely-coupled collection of simple, objective and measurable patterns to save energy. This pattern language (Alexander 1990) not only makes it easier to understand the real issues, but, most important of all, makes it natural to share new knowledge. Thus the real target group for Behavlets are those who are interested in spreading the energy-efficiency message, whether they are developers of ICT solutions or key users.

The Behavlet patterns are intended to be schematic so that the instantiation for a user can take more than one form. Within the SAVE ENERGY project the instantiation takes the form of linked components within the serious game, which helps to bridge the gap between the requirement to address a serious topic and the requirement for some frivolity to help create fun. In terms of the serious game, the main objective of the Behavlets is to address the challenge to close the attitude-behaviour gap between the awareness that energy waste is a problem and behavioural transformation to reduce energy consumption and greenhouse gas emissions.

The Behavlet system is composed of three different levels:

1. Resources, actions and outcomes specified.
2. Elements of (1) assembled in a Behavlet along with contextually-relevant information and an Instrument to help consolidate behaviour transformation.

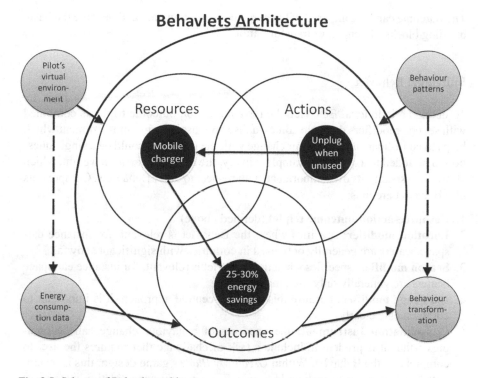

Fig. 8.5 Schema of Behavlet architecture

3. Groups of several relevant Behavlets defined in (2) presented in the context
   of a virtual environment related to the five teams. This is further described in
   Sect. 6.1 below.

Resource—Action—Outcome

The core of a Behavlet is built from discrete components composed of a *resource*,
e.g. some energy-using equipment (e.g. mobile charger), which is linked with an
action (e.g. unplug the mobile charger when not in use vs. leave the mobile charger
plugged in even when not being used) to produce some expected *outcome* (20–25 %
energy savings). This system is shown below in Figure 8.5

To obtain the resource and action definition in the project, the critical incidents
and energy audits in each pilot illustrate some of the conditions of the pilot, namely
building envelope, occupancy and functions, and will point out the mix of resources
on each Behavlet.

In a general situation, on the other hand, the known or expected behaviour patterns from best practices, literature or direct observation can feed the range of resources and actions which users can perform to change the status of the resource.

The outcome can be compared with energy consumption data and become the basic building blocks of behaviour transformation.

Individual Behavlet

To obtain an individual Behavlet, the resource-action-outcome triplet is combined with several modifiers that help contextualise and display, plus an Instrument which helps to consolidate the behaviour change with the user. This could be a single question to be added to a quiz, for example. Finally, the Behavlet carries a weight, which allows it to specify its own importance with respect to other Behavlets. Components of a Behavlet are thus:

1. **Resource-action-outcome triplet** (defined above)
2. **Location modifier:** specifies where the Behavlet is relevant, for instance car-space heaters are generally only used in countries with significant snowfall.
3. **Season modifier:** specifies when the Behavlet is relevant, for instance car-space heaters are generally only used in winter.
4. **Language modifier:** because this is a user-centred approach, it is important to communicate clearly.
5. **Consolidation Instrument:** the concepts of behaviour change can be complex—thus it is prudent to include a task or challenge that requires the user to comprehend the Behavlet. Within *Green My Place's* game design, this is instantiated as a question to include in a quiz. The optional consolidation instrument provides closure for the user, in that once they master this, they can be said to have internalised the Behavlet.
6. **Weight:** it can be useful for development to directly specify the importance of the Behavlet. This can be derived from the outcome.

To summarise, Behavlets are designed to provide a conceptual foundation on which to build a behaviour-changing interaction linking the virtual to the real. In the virtual, the user receives instant feedback from all actions that are explicitly linked to the internal variables, and related by example to external variables in the real. Based on that information, the user can influence the virtual system by changing some of the parameters of his actions. Ongoing real-world monitoring can show the comparable effects of behaviour change in the evolving values of key performance indicators.

5.2   Game Flow

Part of the contextualisation step (the 3rd order) involves transforming game content into something recognisable to the player. In *Green My Place*, this meant that the meta-game and mini-games should reflect each team and the player's own environment or locale. Thus the alignment with context takes place along two axis:

Fig. 8.6 Screenshots from the mini-game WindowWatcher—5 versions for 5 pilots

1. **Social axis:** From *isolated play* to *social interaction*, which defines community and the sense of social norms, which can be a powerful motivator both to play, and to accept and *further disseminate* the serious message (Cowley et al. 2011). In addition to defining teams based around real locations, we also enabled intra-team competition through high scores tables, and inter-player communication via Facebook.
2. **Content axis:** From *generic content* to *situated realistic content*, which defines the nature of the message, and therefore who can receive it. Each game contains variables that in some way attempt to identify it with each team, such as the skin, which can be varied to localise particular elements of the software to a particular team's circumstances.

As an example of the content axis, certain mini-games have, as part of their graphical layout, the capacity to depict a team building. This allows each of the five teams to have its own version of that skin—showing and naming an individual team building.

Additionally, each team is based in a different country, implying language localisation for each one. Thus there will be at least 25 versions of every game, when language and skin variations are combined (although such variations are superficial to game-play). This skin-variation scheme reminds the player that they are playing the mini-game to support *their team*.

Figure 8.6 shows an example of this as it is implemented in the splash screens of the 5 team variants of the mini-game *WindowWatcher*.

6 Modelling (4th order)

The main challenge remaining once this order is reached is *modelling* to support testing (the 4th order)—do the mechanics designed for gameplay support the learning specified for the pedagogy? To some degree this is the level at which existing game design practices would be useful –rapid prototyping and use-case modelling for example.

6.1 Pedagogy Mechanics

Modelling (the 4th order) is when several Behavlets are grouped to create an interactive experience for the user. The reason for this step is that a single Behavlet may not be all that engaging to the user, due to its simplicity. It is non-trivial to *create* the Behavlet, but learning about a simple concept like unplugging a charger can be done very quickly. The problem is that it is counter-productive to learn these concepts quickly, since behaviour change requires a long period of exposure before it is habituated.

To expose the user to a Behavlet over a long period, it is helpful to package several together into an interactive experience. The experience is not just a single engaging product, however, but a product with pre- and post-use extras linking the engagement to real-world data and personal solutions. In the case of the *Green My Place* serious game, the engaging product are the Flash-format mini-games, and the extras are the Learn More containing real-world information, and potentially also a quiz to test the user on all the aspects in the experience.

6.2 Game Mechanics

The interaction of the player with the game was described with a game flow. This equates to an extensive annotated paper design of the interactions taken at the meta-game level, as illustrated by Fig. 8.7.

Once in the mini-game mode the definition of mechanics becomes constrained by the form and duration of play. What remains to specify is how the contextualised elements of the Behavlet architecture can be instantiated in the mini-game, something we discussed at length in Sects. 5.5 to 5.7 of Cowley et al. (2011).

As an example, consider a specific mini-game, namely *Switch Search, as depicted in* Fig. 8.8, a casual puzzle game based on the 'Escape the Room' format. This game instantiates several potential actions as game mechanics, thus representing multiple Behavlets. The story/aim of the game is to find energy wastes, switch them off and thus be able to leave the office for the holidays.

The following are example energy wastes to be found by the player, based around the concept of *waste by neglect* (as mentioned in Sect. 4.2 above), in increasing order of puzzle complexity:

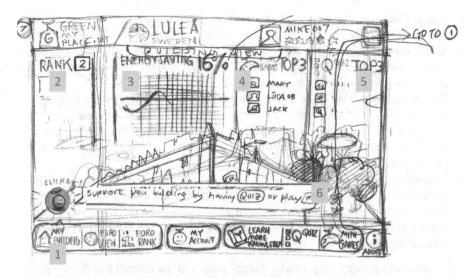

Figure 8.7 Example of the game flow for the team page. Numbers are related to the following key: *1.* (As Guest) Sees a team assigned by the language used, can chose to see any other team thereafter. (As logged user) Sees his/her own team, can chose to see any other team thereafter. *2.* Rank among pilots, based on total score, *which is equal to next three elements* (3.+4.+5.). *3.* Pilot energy savings (also shows information graph on status). *4.* Player contributions (also shows list of top players from pilot). *5.* Achievements of pilot (can be real or virtual/game-based). *6.* Friendly message: *"play games, take quiz or learn more to help improve your pilot"*—three choices corresponding to right-hand section of navbar

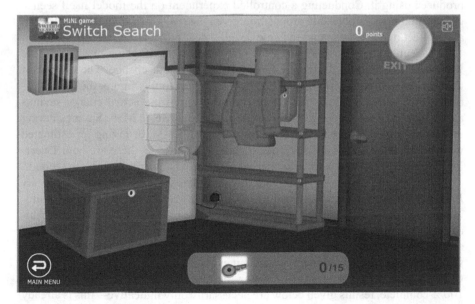

Fig. 8.8 Switch Search. This mini-game encapsulates the waste-by-neglect concept and associated Behavlets regarding stand-by mode, phone chargers and so on

- **Phone Charger**: an open switch object; this is switched on, but there is no phone attached ("vampire power"). The player simply turns it off at the socket switch.
- **Stereo on standby**: an open switch object. The stereo is off, but clearly on stand-by because the blue light is on. Finding and clicking on the hotspot for the power switch completes this switch object's goal.
- **TV on standby**: The TV is off, but clearly on standby because the light is on. However, there is no power switch on the TV itself. The player must find the remote, which is (perhaps) in the crack in the sofa (a hotspot in a sofa item) and can be just barely seen when a cushion cover object is lifted, then click on the power button on the remote.
- **Multiplug Adaptor**: both the TV and the stereo are plugged into an adaptor which is locked inside the TV cabinet. The player need simply open the cabinet doors and turn off the power button on the multi-plug adaptor, however the cabinet is locked. The key can be found in a "junk drawer" of a cabinet that serves as a cover object.

By encapsulating many very similar energy wastes in the overall puzzle, the game forces to player to identify and eliminate, again and again, scenarios which embody the simple concept of waste by neglect.

7 Evaluation

Evaluation of the QUARTIC model is inevitably tied to evaluation of the games produced using it. Conducting a controlled experiment on the model itself seems, to the authors, an implausible proposition given the really massive number of variables affecting each data point (a game development project).

A cogent strategy for measuring the effects of the GMP game must be in line with the SAVE ENERGY strategy for effecting behaviour change. The essence of this strategy is outlined in Figure 8.3 above. Referring to this figure, the accumulation of interventions allows project researchers to plot the actual energy savings against the list of known effects for users throughout M6 to M30. Savings during the period M6-M18 can be offset for the period M18-M30, giving an estimated value to the effect of the serious game on energy saving. As these are both 1 year-long periods, back to back, the first period is directly comparable with the second.

Two forms of evaluation were undertaken to date in the context of *Green My Place*—we assessed the educational value and the energy saved in each pilot building in reference to a baseline. Energy-saved data was recorded by each pilot technical partner using installed equipment within the building, measuring various test objects. Player-related data was obtained from the database of player activity, which was designed to record pertinent facts about everyday use (in addition to supporting game function), such as page visits. Since the timeline for SAVE ENERGY is not 100 % complete, results given below are necessarily only indicative—this is already guaranteed by their correlative nature.

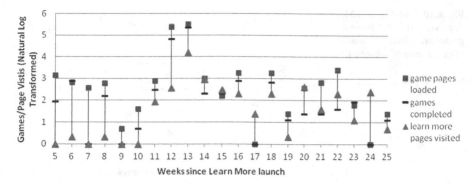

Fig. 8.9 Rate of Learn More visits compared to games played, listed in terms of calendar weeks during 2011

7.1   Educational Value

Analysing the database of player activity in *Green My Place*, we primarily wanted to know whether our pedagogical interaction loop was working or not. The record of game plays and the record of Learn More page visits should provide that, so our metric for education was the rate of Learn More page visits (LMVs) inspired by game plays: see Fig. 8.9.

The number of LMVs was initially low, but as players grew accustomed to the system, LMVs rose (after calendar week 11). Seeing this positive response to the format, we posted some direct links to custom Learn More pages (week 12, followed by more games, and another in week 17). Thereafter, since in every week only one activity is released, some weeks even show high LMVs combined with zero game activity. Comparing LMVs in the weeks of zero game activity against others we see an increase of 23 % when games are played. This is an indication of the effect of game playing on the inclination to learn. Further, by plotting the regression line over the graph of LMVs against games completed (Fig. 8.10), we see that

a. the intercept $y > 0$ confirms that more LMVs occurred without games completed, than *vice versa* (indicating that players more often went to learning content alone, than to game content alone).
b. the slope of the line $m = 0.69$ nevertheless suggests a strong correlation between games completed and LMVs, which we calculate as *correl = 0.58*, $F(1, 19) = 9,6$, $p < .01$.

7.2   Energy Saved

On the topic of energy saved we are one more step removed from the direct impact of the QUARTIC model—however this is the ultimate purpose of the SAVE ENERGY project, so it is appropriate to look at it.

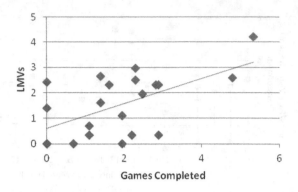

Fig. 8.10 Plot of Learn More page visits (LMVs) against games completed (natural log scale for ease of viewing)

Figure 8.11 shows energy data from three 'test objects' from the Luleå pilot building, a music hall, an office and a restaurant. The project intervened in each of these locations, with meetings, the serious game and energy usage screens in public areas. Baselines were recorded for these test objects in the initial period of the project, and every month from June 2010 until April 2011, the consumption of energy (adjusted for weather variation and estimated number of users) is shown against this fixed baseline value. Thus the savings per month (positive or negative) are shown below also. Comparison of these three test objects tells us a number of things:

- The restaurant is vastly the greatest consumer of energy in the building.
- The music hall is a very variable consumer of energy.
- In relative terms, the office is the greatest saver (has the highest savings value proportionate to consumption).
- In terms of meeting the baseline target, the office is also the most consistent.

This suggests that a difference existed between the office and the other locations. In all three places, the same meetings and energy usage screens were part of the intervention—it is only the presence or absence of personal computers which differs between the office and the others. It is our belief that this difference facilitated the use of the serious game and the better performance in the office reflects it.

8 Conclusions

This paper describes a process model for serious game development, designed to marry methods from entertainment game design to the pedagogical requirements of educational software. Another example of its application can be found in Cowley et al. (2012), while in this paper we examined the case of *Green My Place*, a serious game designed to promote energy efficient behaviour change for the SAVE ENERGY project.

Based on the process model, and with the inclusion of specific additional design models such as the Behavlet, the game was developed to successfully combine

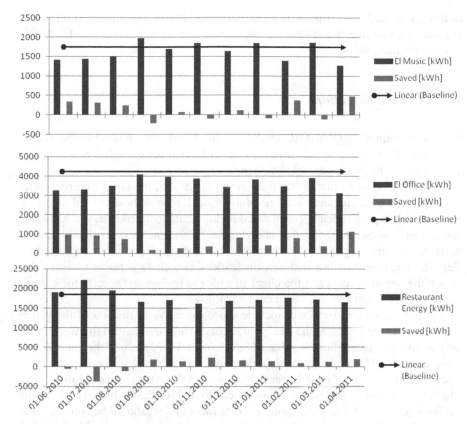

Fig. 8.11 Plot of energy savings in three test objects throughout the Luleå pilot—Music Hall, Office and Restaurant—for the period June 2010 until April 2011. Kilo Watt hours (kWh) consumed is plotted (by *dark grey vertical bars*) against the baseline recorded prior to testing (*horizontal black line*), with the corresponding savings in kWh below (*light grey vertical bars*)

frivolous play and serious teaching. QUARTIC's two streams distinguish between the different foci of the client and the game developers, allowing interactions to ensure that the requirements of the client are met by the game developers. This 'marriage' of game design and player modelling helps to prevent difficult reverse engineering of game mechanics or clunky pedagogy-driven design. The QUARTIC model can facilitate the workflow and communication between people involved in the development of an educational game (or any other TEL-application).

The model may lack specificity over the whole duration of development, since it is meant to be recursive but does not explicitly define how to end one iteration and begin another. The spiral design model which inspired us does do this, and so we note that QUARTIC is only meant to overlay existing software development practices, not replace them entirely. It has been argued that our model strictures the creative process of game design, and this is fair warning in the context of entertainment games—however serious games are, we feel, an area where constraints are definitely needed both to prevent feature-creep and to ensure a minimum level of pedagogical functionality. It has been said that creativity comes more readily within constraints than without.

8.1 Future Directions

By understanding the QUARTIC model and the terms used to describe the stages, communication regarding the current status in both streams might become easier. Additionally, the process model shown in

Figure 8.1 could be literally used to provide a visualization of the current status in the project, enforcing the discipline of the model. It could be also used for an agile approach to project management and development, complementary to workflow models such as Scrum (Schwaber and Beedle 2002). Each stream should have at least one 'leader' responsible for the workflow and progress in their own streams. Periodic meetings between both stream leaders as well as periodic meetings between the stream leaders and the client in order to update and specify the client`s requirements would enhance the overall workflow.

Further work should and will include more extensive evaluation of the games created. Over the course of the next year, experiments are planned to ascertain the educational and entertainment value of both *Green My Place*, and the game developed in the TARGET project (which also used the QUARTIC process). By inference from the results of these experiments, we can draw some conclusions regarding the model itself. A controlled experiment to test the model directly would not be feasible: given that any one game development project should be considered as only a single data point and several test and control data points would be needed, such a large study is unlikely. However, we hope that if the model is used again in independent projects, these case studies would allow more evaluation. Whether it is used by others or not, we urge those working in the serious games space to pay more attention to resolving this quite unique demand to unify the respective expertise of game designers and pedagogy designers.

Acknowledgements I would like to thank Chris Bateman for advice on the design of *Green My Place* and considerable feedback on drafts of this paper; Jose Moutinho for collaboration in developing the Behavlet concept; and Tuija Heikura for helping to conceive of the QUARTIC process model. Additionally, Aki Jäntti, Zhang Jiongkai, Eva Szadeczky-Kardoss and Heikki Silanpää all contributed to the development of the components of *Green My Place* pictured above.

Part of this work was supported by the European Union through the SAVE ENERGY project (CIP-ICT-PSP-238882).

References

Abrahamse, W., Steg, L., Vlek, C., & Rothengatter, T. (2005). A review of intervention studies aimed at household energy conservation. *Journal of Environmental Psychology, 25*(3), 273–291.

Abrahamse, W., Steg, L., Vlek, C., & Rothengatter, T. (2007). The effect of tailored information, goal setting, and tailored feedback on household energy use, energy-related behaviors, and behavioral antecedents. *Journal of Environmental Psychology, 27*(4), 265–276.

Alexander, C. (1990). *A pattern language*. München: Fachhochsch., Fachbereich Architektur.

Bateman, C., & Boon, R. (2005). *21st century game design* (Vol. 1st). London: Charles River Media.

Boehm, B. W. (1995). A Spiral Model of Software Development and Enhancement. *IEEE engineering management review., 23*(4), 69.

Carpenter, A. & Windsor, H. (2011). Ahead of the Game? – Games in Education. Retrieved from http://seriousgamessource.com/features/feature_061306_ahead_of_the_game.php

Charles, D., & Black, M. (2004). *Dynamic Player Modelling: A Framework for Player-centred Digital Games*. Paper presented at the Proceedings of 5th International Conference on Computer Games: Artificial Intelligence, Design and Education (CGAIDE'04).

Cook, D. (2006). The Chemistry Of Game Design. *Gamasutra*. Retrieved from http://www.gamasutra.com/view/feature/1524/the_chemistry_of_game_design.php

Cowley, B., Charles, D., Black, M., & Hickey, R. (2006). *User-System-Experience Model for User Centered Design in Computer Games*. Paper presented at the Adaptive Hypermedia and Adaptive Web-Based Systems.

Cowley, B., Moutinho, J., Bateman, C., & Oliveira, A. (2011). Learning Principles and Interaction Design for 'Green My Place': a Massively Multiplayer Serious Game. *Entertainment Computing, 2*(2), 10.

Cowley, B., Bedek, M., Heikura, T., Ribicro, C., & Petersen, S. (2012). The QUARTIC Process Model to Support Serious Games Development for Contextualized Competence-Based Learning and Assessment. In M.-M. Cruz-Cunha (Ed.), *Handbook of Research on Serious Games as Educational, Business and Research Tools: Development and Design* (Vol. (in press)). New York: IGI Global.

Darby, S. (2006). *The Effectiveness Of Feedback On Energy Consumption. A Review for DEFRA of the Literature on Metering, Billing and direct Displays*. Oxford: University of Oxford.

Egenfeldt-Nielsen, S. (2006). Overview of research on the educational use of video games. *Nordic Journal of Digital Literacy, 3*, 23.

Ehrhardt-Martinez, K. (2008). *Behavior, energy, and climate change policy directions, program innovations, and research paths*. Washington, D.C.

Espejo, R., & Harnden, R. (1989). *The Viable system model: interpretations and applications of Stafford Beer's vsm*. Chichester, West Sussex, England; New York: J. Wiley.

Gee, J. P. (2003). *What Video Games Have to Teach Us about Learning and Literacy*. New York: Palgrave Macmillan.

Gee, J. P. (2006). Are Video Games Good for Learning? *Nordic Journal of Digital Literacy, 3*, 10.

Goldberg, L. R. (1993). The structure of phenotypic personality traits. *The American Psychologist, 48*(1), 26–34.

Karweit, D. (1993). Contextual learning: A review and synthesis. Baltimore, MD: Johns Hopkins.

Kolb, D., A. (1984). *Experiential Learning: Experience as the Source of Learning and Development* Beverley Hills: Sage Publications.

Koster, R. (2005). *A theory of fun for game design*. Scottsdale, AZ: Paraglyph Press.

Lave, J., & Wenger, E. (1991). *Situated learning: legitimate peripheral participation*. Cambridge [England]; New York: Cambridge University Press.

Lazzaro, N. (2009). *Creating an MSO: Viral Emotions and the Keys to Social Play*. Paper presented at the Game Developers Conference. Retrieved from http://www.slideshare.net/NicoleLazzaro/gdc09-mso-slides-100n032609

Lockwood, M., & Platt, R. (2009). *Green Streets*. United Kingdom: Institute for Public Policy Research.

Marsh, T., Yang, K., & Shahabi, C. (2006). *Game development for experience through staying there*. Paper presented at the Proceedings of the 2006 ACM SIGGRAPH symposium on Videogames.

McGinnis, T., Bustard, D. W., Black, M., & Charles, D. (2008). *Enhancing E-Learning Engagement Using Design Patterns from Computer Games*. Paper presented at the Proceedings of the First International Conference on Advances in Computer-Human Interaction.

Mislevy, R. J., Steinberg, L. S., Almond, R. G. & Lukas, J. F. (2006). Concepts, terminology and basic models of evidence-centered design. In D. M. Williamson, R. J. Mislevy, and I. I. Bejar (Eds.): Automated Scoring of Complex Tasks in Computer-Based Testing (pp. 15–47). Mahwah, N. J.: Lawrence Erlbaum Associates

Moutinho, J. L., Cowley, B., & Oliveira, A. (2010). *Behavlets: A Pattern Language for Energy Efficiency in Public Buildings*. Paper presented at the BEHAVIOR, ENERGY & CLIMATE CHANGE.

Oliveira, A., Fradinho, E., & Caires, R. (2006). *From a Successful Regional Information Society Strategy to an Advanced Living Lab in Mobile Technologies and Services*. Paper presented at the International Conference on System Sciences.

Petersen, J. E., Shunturov, V., Janda, K., Platt, G., & Weinberger, K. (2007). Dormitory residents reduce electricity consumption when exposed to real-time visual feedback and incentives. *International Journal of Sustainability in Higher Education, 8*(1), 18.

Raybourn, E. M. (2007). Applying simulation experience design methods to creating serious game-based adaptive training systems. *Interacting with Computers, 19*(2), 206–214.

Schell, J. (2008). *The art of game design: a book of lenses*. Amsterdam; Boston: Elsevier/Morgan Kaufmann.

Schwaber, K., & Beedle, M. (2002). *Agile software development with Scrum*. Upper Saddle River, NJ: Prentice Hall.

Squier, L. (2001). *What is Data Mining?* Reston, VA: Data Management Association National Capital Region.

Squire, K. (2006). From Content to Context: Videogames as Designed Experience. *Educational Researcher, 35*(8), 19–29.

Wong, W.-L., Shen, C., Nocera, L., Carriazo, E., Tang, F., Bugga, S., et al. (2007). *Serious video game effectiveness*. Paper presented at the Proceedings of the international conference on Advances in computer entertainment technology.

Chapter 9
3-D Manufacturing: The Beginning of Common Creativity Revolution

Robert Niewiadomski and Dennis Anderson

We shape our tools and thereafter our tools shape us.
Marshall McLuhan

1 The Common Creativity

Leonardo da Vinci's extraordinary creative versatility became the ultimate embodiments of the Renaissance humanist ideal. His codices contain numerous astonishingly futuristic blueprints of scientific and engineering inventions, among them flying machines (see Fig. 9.1) and hydraulic pumps. The staggering design of these inventions exceeded the available technology of the time. The limitations of the Renaissance manufacturing processes and materials prevented many of these inventions from reaching even the stage of a prototype.

Another iconic figure of the Renaissance, Johannes Gutenberg, is credited with the introduction of a device that led to a dramatic technological transformation. The invention of the printing press in 1450 ushered in an unprecedented dissemination of literacy and knowledge (see Fig. 9.2). Thanks to the press, the relatively unrestricted circulation of ideas broke the information monopoly of the Church and the aristocracy.

In the absence of intellectual property rights and prior to the emergence of the modern concept of authorship, subversive literature containing a cornucopia of ideas spread across Europe and evolved through continual adaptations and mutations. The Gutenberg's press played a key role in the propagation of ideas of Luther, Copernicus, Newton and Lock. It created the conditions for intellectual ferment and it ultimately led to the Reformation, Scientific Revolution and the Enlightenment. The inception of common creativity had arrived and enabled many to

D. Anderson (✉)
St. Francis College, Brooklyn, USA
e-mail: danderson@sfc.edu; dennis.danderson@gmail.com

R. Niewiadomski
New York City Department of Education (NYCDOE) and Teach For America, New York, USA
e-mail: rn2233@gmail.com

N. Lee (ed.), *Digital Da Vinci*, DOI 10.1007/978-1-4939-0965-0_9,
© Springer Science+Business Media New York 2014

Fig. 9.1 Many of Leonardo's drawings reflect his fascination with the phenomenon of flight. This particular sketch depicts a machine resembling a helicopter. (Source: ausschnitt aus http://commons.wikimedia.org/wiki/Image:Leonardo_da_Vinci_helicopter_and_lifting_wing.jpg)

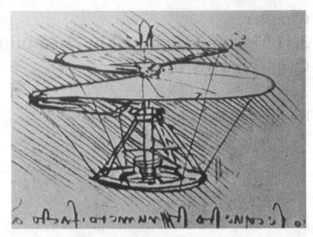

Fig. 9.2 The introduction of the movable type printing by Gutenberg in 1450 initiated the process unprecedented dissimilation of information and knowledge. (Source: http://en.wikipedia.org/wiki/File:Printer_in_1568-ce.png Meggs, Philip B. *A History of Graphic Design.* John Wiley & Sons, Inc. 1998, p. 64)

tap their imaginative powers to the genius of the few. Their visions proliferated and evolved laying the foundations for present modern societies.

Fig. 9.3 Additive technology turns digital blueprints into three-dimensional objects. (Source: http://3dprinting-al-alwi.blogspot.com/)

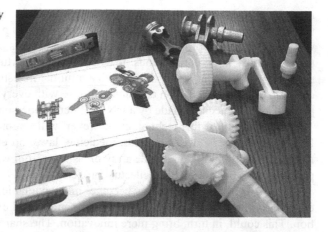

Common creativity has a chance to emerge when people's ingenuity is matched by a widely accessible technological vehicle capable of its distribution. This, in turn, may open ideas to mutations and expedite creativity through the communal input. There were a various printing techniques in existence in China and Egypt prior to Renaissance, however, their impact was limited to their own societies. As oppose to Gutenberg's press, they did not lead to explosion and spread of ideas. One might wonder why? Was the lack of their mass accessibility responsible? Perhaps homogeneity and centralized power structure of these societies was the one of the contributing factors?

Today, a distant successor of the Gutenberg invention could empower individuals who share Da Vinci's revolutionary imagination to realize his ideas. The additive manufacturing technology also known as 3-D printing can ease the process of turning ideas into testable models as well as complete products. If Da Vinci were alive today, what would he do with this remarkable technology?

In this chapter, we will explore the potential of 3-D printing to change the economic paradigm and challenge the traditional distinction between producers and consumers. We will discuss how this emerging technology might affect the established business model and transform the current manufacturing landscape. Will digital manufacturing replace the established production paradigm or merely complement it?

The rapidly growing applications of additive manufacturing are both exciting and astounding. Highly customized products made with 3-D printers are already available on the market (see Fig. 9.3). Other projects are reaching promising stages of development: doctors are manufacturing 3-D prosthetics, experimenting with printing pharmaceuticals, human tissue and even organs. A company called *Made in Space* will soon help astronauts print out parts such as wrenches and nails for spacecraft and space stations while in orbit. A Dutch architect launched a project to 3-D print an entire house (Crook n. d.). Is this invention a dawn of a new technological revolution, as President Obama proclaiming in his 2013 State of the Union?

Chris Anderson, the former editor of *Wired* who's *Makers* helped bring the growing 3-D printing economy to light, believes that the introduction of a digital manufacturing model to the general public will undoubtedly lead the democratization of technology and it will accelerate the technological progress. Individual users will soon be able to produce small numbers of products comparable in quality to those made by large corporations. It will provide opportunities to make inexpensive prototypes, manufacture things with a minimal investment, share, modify, and customize designs. Anderson envisions a future with everything from custom body parts to instant vaccines made by a DNA printer (Anderson 2012).

The versatility of this technology, however, will undoubtedly pose many challenges to various fields: economics, medicine, law, and ethics. It will force us to reevaluate our thinking on a wide array of issues—from what we consider morally permissible to our notion of intellectual property.

As the 3-D printing technology becomes more available, the community of makers grows. They are sharing designs, experimenting and expressing their imagination. This could, in turn, bring more innovation. The shared creativity that already thrives on the Web, has finally met its physical counterpart soon available virtually to anyone. The selective creativity and innovation that happened at the industrial level will move down to the grassroots. Should production become a mundane process, what people will truly value will be the ideas themselves. Their commoditization will transform them into a type of modern currency.

2   What is Additive Manufacturing?

The additive technology used in 3-D printing has been available for over 30 years but only quite recently gained a considerable publicity and captured the imagination of both that industry as well as the media. Charles Hull is credited with the invention and patenting of the first effective 3-D printer in 1986. Hull's printer used ultraviolet to cure successive layers of liquid material deposited on top of each other (U.S. Patent 4,575,330).

The early 3-D printers were large, expensive, and had limited applications. Thus, at that time, there were not considered to be a viable option to the established manufacturing process. This perception has shifted when the 3-D printing overcame technological limitations and the printers themselves became more affordable.

How does the 3-D printing work? In the first step, an object can be scanned, then the software translates it into a digital file or one can download a virtual blueprint for a computer aided design (CAD). Next, the virtual blueprint is "sliced" into digital cross-sections. Subsequently, the printer reads these cross-sections and replicates them as layers of specialized material each one printed directly on top of the previous one (Barnatt 2013, p. 4). The process is complete when the whole three-dimensional object has been "printed" with these layers.

The typical raw material use for the process is a powder, most commonly -a thermo polymer or a metal (Excell and Nathan 2010). Depending on the material used, the ways the printer deposits it in layers can vary. Some printers have a blade mounted on a moving arm that sweeps an even layer of the powder on top of the work surface inside the chamber. Next, it scans back and forth over the surface, melting the powder in the shape of the first layer. The work surface then moves down by the thickness of the layer and another layer of powder is distributed over the surface. Other 3-D printers utilize electron beams instead of lasers. They have the advantage of transmitting more energy and thus they can melt the powder faster. The final product, however, typically has a rougher surface finish that requires additional treatment.

Since different materials are suitable for different purposes, some objects require more than one kind of material. Some 3-D printers already reached this level of sophistication and can simultaneously use different materials to create complex objects. Currently, they are out of reach to a home user but this might soon change. Will they become a common household commodity such as a desktop computer? (See Fig. 9.4). Ultimately, the quality of 3-D printers and their availability are the critical aspects that will determine whether individual home inventors become manufacturers.

3   The Third Industrial Revolution?

Major technological revolutions transformed the landscape of what kinds of goods are made, where and how they are made, and who makes them. It appears that we are currently in the midst of such transformation. Technologically advanced and wealthy nations struggle to solve the problem of worldwide unemployment. The massive disappearance of jobs in manufacturing, agriculture and service industries can only be partially explained by the current economic crisis. Perhaps most of these jobs will never come back for entirely different reasons. Efficiency no longer requires a legion of workers and economic growth doesn't necessarily translate into job creation. Jeremy Rifkin prophesized in 1995, in his controversial book *The End of Work*, that automation will ultimately have a devastating impact on blue-collar, retail and wholesale workers (Rifkin 1995).

Today, the application of advanced software eliminates jobs in areas that previously required considerable training and knowledge. Work as we know it is dying. In feudal economy, labor used to be the burden of the many. In modern times, we tend to think about it as our right. Soon, however, work might become the privilege of the few. Thus, we ought to fundamentally rethink the future functioning of a society where only the minority might enjoy the privilege of full-time secure employment.

The first industrial revolution started in the late 1700 in Britain with the invention of the steam engine and the subsequent mechanization of the textile industry (see Fig. 9.5). The cotton mill replaced hundreds of weaver's cottages. The birth of a factory eliminated these in-home laborers but also created new opportunity and ushered in development of urban centers. Henry Ford's application of the assembly line signified the beginning of second industrial revolution and gave birth to mass

Fig. 9.4 The key to the
future success of 3-D printing
and its impact on society
lies in their availability.
Will 3-D printers become as
common as PCs? (Source:
http://en.wikipedia.org/wiki/
File:MakerBot_ThingO-
Matic_Bre_Pettis.jpg)

production. The more current rapid expansion of automation eliminated the need for
massive human work force.

As the traditional model of manufacturing is approaching its agony, there is a
sense of urgency to find a new paradigm. Rifkin proclaims the arrival of the third industrial revolution in which 3-D printing will play a critical role. His new economic
narrative emphasizes the shift from the hierarchical order of controlling energy,
information and manufacturing towards a lateral, cooperative, and nodal model.

The great economic revolutions in history, he argues, occur when new communication technologies converge with new energy systems: "In the nineteenth century, cheap steam powered print technology and the introduction of public schools
gave rise to a print-literate work force with the communication skills to manage
the increased flow of commercial activity made possible by coal and steam power
technology, ushering in the First Industrial Revolution. In the twentieth century,
centralized electricity communication -the telephone, and later radio and television
-became the communication medium to manage a more complex and dispersed oil,
auto, and suburban era, and the mass consumer culture of the Second Industrial
Revolution. Today, Internet technology and renewable energies are beginning to

Fig. 9.5 The first industrial revolution centralized production and ushered in the rapid development of urban centers. (Source: http://en.wikipedia.org/wiki/File:Hartmann Maschinenhalle_1868_%2801%29.jpg)

merge to create a new infrastructure for a Third Industrial Revolution (TIR) that will change the way power is distributed in the twenty-first century. In the coming era, hundreds of millions of people will produce their own renewable energy in their homes, offices, and factories and share green electricity with each other in an *Energy Internet* just like we now generate and share information online (Rifkin 2013)."

Radical democratization of communication and manufacturing are the key factors contributing to the third industrial revolution. We can already observe the impact of the Internet on the music industry and publishing field where the "democratization of communications has enabled nearly one third of the human population on earth to share music, knowledge, news and social life on an open playing field, marking one of the great evolutionary advances in the history of our species (Rifkin 2013)."

Rafkin predicts that similar drastic change of the business model will occur in manufacturing thanks to 3-D printing. The third industrial revolution, as he clam, will empower people not only to produce their own virtual information and energy but also the customized production of goods. Soon, "millions of people might start making batches or even single manufactured items in their own homes or businesses, cheaper, quicker, and with the same quality control as the most advanced state-of-the-art factories on earth (Rifkin 2013)."

The innovative spirit of the Steve Jobs' generation empowered a vast portion of human population to share information, ideas and fruits of their creativity causing what Rifkin describes as a "fundamental change in the way capitalism functions that is now unfolding across the economy and re-shaping how companies conduct business (Rifkin 2013)." Online blogging, electronic publishing and file sharing dramatically reduced transactions costs, the speed of idea dissemination and, perhaps even more importantly, the scope of depth of public participation.

We can easily see the resemblance between the impact of personal computers on music, media and publishing industry and the possible impact of 3-D printing on traditional manufacturing. Rifkin predicts that manufacturing landscape will be radically transformed: "Expect similar disruptive impacts as the diminishing transaction costs of green energy allow manufacturers, service industries, and retailers to produce and share goods and services in vast economic networks with very little outlay of financial capital (Rifkin 2013)."

Thus, the arrival of this new technology forces us to reevaluate the entire way we think of industrial production. The highly capitalized, massive and highly capitalized factory staffed with blue collar workers and assembly lines may no longer be dominant element of the production model. Provided that inexpensive and technologically advanced 3-D printers become widely available, we might be soon manufacturing customized items in our homes or small businesses.

The first and second industrial revolution promoted a business model that relied on companies with vast financial resources capable of financial obstacles associated with designing, prototyping, actual manufacturing and finally marketing a new product. Today, the Internet already reduces the entry costs in generating and disseminating information. The marketing no longer requires the services of centralized communication including printed publications radio and television.

Additive manufacturing will transform this traditional business model even further. We can imagine that widely available 3-D printing could respond much faster than traditional manufacturers to even the most whimsical individual customers' needs and exotic, unexplored niches. Thus, a sufficiently great numbers of such enterprises can have a cumulative effect challenging and potentially out-competing established giant manufacturing companies. Does that mean that global companies will disappear completely? Not necessarily. Rifkin argues that they will "increasingly metamorphose from primary producers and distributors to aggregators. In the new economic era, their role will be to coordinate and manage the multiple networks that move commerce and trade across the value chain (Rifkin 2013)."

Moreover, 3-D printing might encourage individuals to re-design, modified and customized main stream products for commercial or personal use on a much larger scaled than currently practiced. Depending on the scale of this phenomenon, the current system of enforcing patent and trademark rights might be pushed to its limits. Perhaps the very concept of patents itself will have to be fundamentally reconceived altogether.

Democratization and customization of manufacturing, although important, are not the only features of the emerging revolution. Perhaps the change that matters the most will have to do with the accuracy, speed and affordability of rapid prototyping allowed by 3-D printing.

Currently, there appears to be an increasing recognition in the political circles of the inevitable change of the established manufacturing model. The Obama

Fig. 9.6 A cutting edge application of additive manufacturing: scientists are experimenting with 3-D printing human tissue and body parts. (Source: 3D printer ears http://news.cornell.edu/stories/2013/02/bioengineers-physicians-3-d-print-ears-look-act-real)

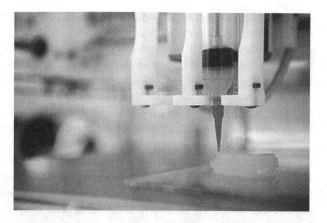

administration took an important step to ensure that the US remains a leader in the field when it created the National Additive Manufacturing Innovation Institute, a public-private partnership dedicated to advancing 3-D printing. Its goal is to help train the workforce needed in this growing field, develop curriculums at technical schools, offer sites and equipment where businesses can validate ideas, and support research that will let domestic suppliers produce the advanced machinery the industry will need.

4 Print Out Your Thoughts

What can this technology do? Can it really spark the common creativity revolution as previously asserted? How can those that are not digitally savvy get involve and benefit from 3-D printing?

According to Chris Anderson, 3-D prototyping will have a massive impact on our society. It will prompt the *maker revolution*, as he calls it, which will rival previous major technological shifts -the industrial revolution and the widespread of PCs. Anderson believes that the innovative products that will someday come from affordable 3D printers will be limited only by the human imagination (Anderson 2012).

In fact, rapid prototyping spurred by digital manufacturing has been already embraced by large businesses. This resulted in a significant improvement in the product-development process across a wide array of industries, including the manufacturing of cars, consumer electronics, safety equipment and medical devices. 3D printing is an inexpensive way to crate and test multiple visions of the prototype. Tested version can be quickly updated and successive duplications of a new product can be refined before being mass-produced. The result is more innovative, higher-quality products.

Companies are no longer constrained to software modeling to predict the performance of their new product. Although virtual prototyping has many advantages, it is somewhat limited. Having a physical model is invaluable when it comes to assessing the actual performance of a new product.

The success of the 3-D revolution partially relies on how successful the companies are not only in making both the software and the printers accessible but also how easy it is to operate. In addition, the public has to be educated in basic literacy of this innovation. As with desktop computers, the widespread of the technology and time will partially take care of the problem. Most young children currently in the developed countries are computer literate even without formal training. We may predict they will rapidly absorb the new technology. Adults may need a little more assistance on the other hand. Some companies already offer classes to demonstrate the possibilities of 3-D printing. For example, TechShop, a technology incubator with offices in several U.S. cities, offers affordable workshops where just about anyone can learn how to prep software and print simple objects such as nuts and bolts.

Although touted possibilities of additive technology appear to be endless, the design software is not easy to master, not only for those less design inclined but also to children that haven't yet learned how to hold a pencil properly. How can we open this field of creativity to those that don't lack imagination but haven't mastered the skills to express it?

The engineers at *Thinker Thing*, the Chilean start-up, claim that technical skills won't be necessary to design objects prior to printing them. They developed a mind-controlled 3-D printing system that determines the shapes of the object according to the wishes of its designer, as gleaned from the headset picking up his brainwaves (Ruz 2013).

Realizing the limitation of the new technology inspired Bryan Salt, the CEO of Thinker Thing: " What is the point of these printers if my son cannot design his own toy? (Ruz 2013)" He stumbled upon in important point—although it appears that 3-D printers could help unleash our inner creativity, freeing us from the constraints of traditional production methods, in reality, those unwilling or unable to plough through the hurdles of instructional manuals could be limited to downloading only ready-made models designed by someone else.

This desire to make the creative process truly democratic, sparked the emergence of what became known as the Emotional Evolutionary Design (EDD)—the software that can interpret user's thoughts. The software is currently used in the Monster Dreamer Project launched by *Thinkers Thing*. The software creators claim that the user will sit in front t of a computer running Monster Dreamer and will be presented with a series of different shapes. This is based on the notion that most people are better at critiquing a design, evaluating whether they like or dislike the shape, than they are at creating new ideas from scratch.. This seems to be a plausible assumption since most innovations are actual modifications of previous designs. This seems to be the way creativity works. The shapes on the screen will mutate randomly but the program will prevent them from becoming too abstract. A headset will pick up the user's emotional reaction to these shapes with a build-in electroencephalograph. The device used by the company has sensors on the scalp that register different patterns of brain activity correlated with certain states such as excitement or boredom. Based on these patterns, the computer will identify shapes on the screen that are associated with positive emotional responses. The desirable shapes will then grow bigger on the screen while the others shrink. Subsequently, the biggest part is combined to form body parts and the process continues until the model is completed and ready for printing as a solid object (Ruz 2013).

The premise of the EDD is that anyone can create something without any particular set of technical skills. Its significance to fuelling the common creativity cannot be underestimated. As mention before, one of the concerns is that the general public might have access to 3-D printing but might not be able to do much with it because of the relative sophistication of the available software. If successful, the EDD combined with additive technology might potentially put the means of creation and production the hands of every person.

5   Print Out Your body

The Renaissance period was marked by immense interest in human body (see Figs. 9.7 and 9.8). Artists strived to achieve more realistic and refined portrayal of the human anatomy. Da Vinci and Michelangelo resorted to prohibited then dissections to gain a firsthand knowledge. The dissections they undertook at various stages of their career set the Renaissance standard for anatomical mastery. Later on, Descartes suggested the notion that the human body is a complicated machine. Originally construed as a component of the body-mind dualism, the notion became an important element of the Western philosophy. Finally, physicalism responded to the Cartesian body-mind dualism by asserting that the mind itself is merely the function of the brain.

This framework of body-as-a-machine was adopted in Western medicine. If the body is a type of a sophisticated machine, the reasoning seems to go, it consists of parts that can be improved, fixed or even replaced. However, it was technically quite difficult to manufacture highly customized prosthetics; not to mention that actual functioning organs for transplants have to be harvested from other bodies. This is where additive technology comes into play and is starting to revolutionize the medical field.

Currently, there are two basic application of additive technology in the medical field. One is becoming quite common and it has to do with using 3-D printers to make highly customized and less expensive prosthetics. Second, just emerging, pertains to experimenting with additive technology to create living tissue and functioning organs. The latter one is sometimes referred to as bio-printing or bio-manufacturing.

Some patients already use hip replacements, dental crowns or even cranial implants that have been entirely produced by using digital manufacturing. Not surprisingly, the largest-volume application of additive manufacturing is in the production fairly uncomplicated customized hearing-aid casings. They are now almost exclusively made using additive techniques.

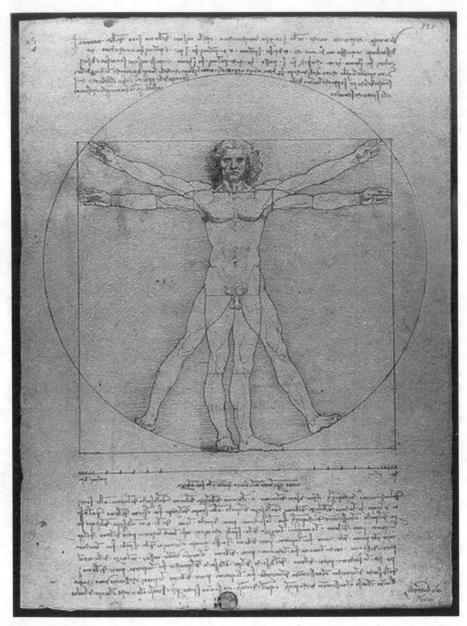

Fig. 9.7 The Renaissance artists displayed a deep interest in human body. (Source: Wikipedia. com, File:Da Vinci Vitruve Luc Viatour.jpg, http://en.wikipedia.org/wiki/File:Da_Vinci_Vitruve_ Luc_Viatour.jpg)

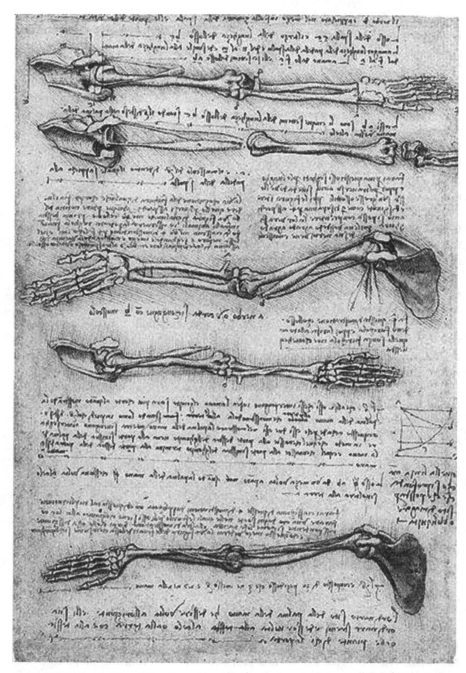

Fig. 9.8 Performing dissections allowed Da Vinci to explore the inner machinery of the human body. (Source: http://en.wikipedia.org/wiki/File:Studies_of_the_Arm_showing_the_Movements_made_by_the_Biceps.jpg)

Curiously though, it appears the biggest drive for adoption of the digital manufacturing in medicine is not customization, but actually cost-effectiveness. For instance, sockets of the hip joint (acetabular cups) manufactured by using 3-D printers have several advantages over the ones made by traditional methods, financial being a very important one (Excell and Nathan 2010). The caps have to have a porous surface that allows the bone to grow into and bond with the implant. The traditional manufacturing process required a depositing an additional layer of porous material onto already made cup. for an acetabular cup (the socket of the hip joint) is a drop forging that is CNC machined and then has a coating put on it for the bone to grow in to the coating. With additive technology, the manufacturer can control the density of the deposited material thus achieving the desired sponginess without additional, and costly, operations.

Individuals can quite literary take things into their hands and manufacture, by using 3-D printing and MakerBot, cheap plastic replacements for their simple prosthetics. They might not last long in comparison to their metal counterparts from medical companies but the amount of money saved is massive. In addition, one might easily make a new one on demand whenever needed. This especially makes sense when prosthetics have to be often adjusted due to body growth.

Some of the cases of customized prosthetics are quite extraordinary. An eighty three-year-old British woman suffered from chronic bone infections. A reconstructive surgery was needed but doctors worried that due to her advanced age such drastic procedure would likely caused complications. Instead they decided to manufacture a custom made transplant of the lower jaw by using additive technology. The implant was created by metal part maker LayerWise. The company printed the 3-D model with titanium powder that was heated and infused together in layers. The whole metal jaw took only a few hours to print and because it was made to fit this particular patient, the surgery was a success and the recovery time-short. Doctors involved with the surgery believe that surgeries involving 3-D printed custom made prosthetics will eventually become routine ("Transplant jaw made," 2012).

Another spectacular example involves a twenty-month old baby boy Kaiba that suffered from a severe version of tracheobronchomalacia. The condition caused his bronchus to collapse. Doctors at the C.S. Mott Children's Hospital turned to digital imaging and scanned boy's malfunctioning organ. Then, they used a 3-D printer to make one hundred tiny tubes which were subsequently laser-stitched them together over the baby's trachea. The tubes are made out material that will eventually dissolve in the boy's body. By then, his trachea is expected to grow into a healthy state (Marchione 2013).

Printing prosthetics is not limited to replacing human body parts only. Veterinarians use the same technology while treating animals. CNN reported a case of a duck from the Feathered Angels Waterfowl Sanctuary in Tennessee that was born with a backwards left foot. That painful and prone to infections foot was amputated and replaced with a 3-D manufactured prosthetic. (Aamoth 2013).

Doctors are also applying additive technology to print models of organs or pathological growths prior to performing complicated surgeries. They can take a CAT scan and literarily print out what, for example, a tumor looks like before they per-

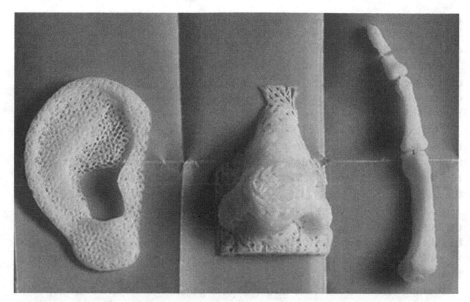

Fig. 9.9 Scientists are using 3D printers to manufacture scaffolds that can be coated with cells and become body parts. (Source: Smithsonianmag.com, "What Lies Ahead for 3D Pringing," http://www.smithsonianmag.com/science-nature/What-Lies-Ahead-for-3-D-Printing-204136931.html?onsite_source=smithsonianmag.com&onsite_medium=internal&onsite_campaign=photogalleries&onsite_content=What%20Lies%20Ahead%20for%203-D%20Printing?)

form the actual surgery. The printed model can be touched and examined prior to any surgical intervention. This can greatly improve the precision and ultimately the success of the actual surgery.

Printing customized prosthetics might be somewhat impressive but hardly paradigm-changing. Scientists are also experimenting with printing actual biological tissue and organs and that might have consequences that we cannot quite foresee, both for medical field as well as our lives (see Fig. 9.9). How would this affect our lives if the tens of thousands of people waiting for organ transplants in the United States alone didn't have to wait? What if victims burned in an accident could replace their scars with skin that was identical to their own? What if an amputee could get back a limb that felt, looked and behaved exactly as the original?

The nascent nature of the field is reflected by the lack of solidified nomenclature: in addition to the mention "bio printing", it is also called "organ printing", "computer-aided" tissue engineering or "bio-manufacturing". In a nutshell, the goal of this developing field is to use additive technology to actually print living human functional tissue as well as living three-dimensional entire organs.

Organ printing is an incredibly complex and challenging task. The technique involves a type of 3-D printer with a cartridge filled with cells and a crosslinker. The cartridge moves back and forth petri dish filled with liquid. The crosslinker is a chemical that causes the liquid in the petri dish to gel, giving the printer a soft jellified surface to print the cells on. The process can be repeated over and over, adding liquid, jellying it, printing more cells, and building layer upon layer, eventually creating a three dimensional object.

Surgeon Anthony Atala and his team form Wake Forrest Institute for Regenerative Medicine are working on engineering different types of tissues and whole organs using additive technology. The team already engineered the first lab-grown bladder. (Atala 2012).

The technique, however, is still facing many hurdles. In every organ, there are blood vessels feeding the organ to keep it alive and working properly. While printing tissue that is lacking blood vessels, scientists are limited to a maximum of about two inches of thickness. Otherwise, when the bloodless printed tissue is too thick, the cells on the inside will die. Crossing that threshold presents one of the technique's first big obstacles. Hence, the problem facing researchers in building an organ for use in a human: How do you get the printed organ to grow and maintain blood vessels? In order to solve this problem the scientists would have to find a way to print vessel-like channels inside the manufactured tissue that would allow driving blood into the organ.

Printing blood vessel inside manufacture organs might require more sophisticated 3-D printers but scientists are optimistic that this technical problem will be eventually solved and when it is, the complex internal structures of organic materials could be reproduced. Achieving this threshold would perhaps mean the end of problems associated with organ harvesting, abuses related to its perpetual scarcity and patients won't have to worry about rejection. After all, the replacement parts would be catered to the individual receiving it.

6 Guns and Drugs

New ethical dilemmas often arise with successive stages of the evolution of civilization. Technological advances open fields of experimentation that bring about consequences that sometimes lead to ethical anxieties. These consequences are not always necessarily socially detrimental; sometimes we are prone to condemn an emerging phenomenon because we lack an adequate ethical framework to process and examine it. This was true about organ transplant, cloning, in vitro, and the Internet. It is also certainly true for 3-D printing. As the technology constantly evolves, so must our notion of what is morally permissible.

Why then would 3-D printing be the cause of ethical concerns? When people can replicate any object with ease, the worry is that there are plenty of objects we typically would not want to be replicated. We might think that we have good reasons for strict control of such replication. Media reported that a German group used his 3-D printer to create a key to unlock handcuffs carried by the Dutch police. Startlingly, they were able to reproduce the key accurately by using nothing more than a photograph of the key hanging from the belt of a police. They applied some simple math to estimate the size of the key. In addition, they put its digital model up online for anyone to print (Greenberg 2012).

Eventually, it turned up that the key was just a proof of concept by an enthusiastic amateur and hasn't been used in the commission of any actual crimes. Obviously, printing keys by using only photographs might be seen as a curious nuisance but can hardly gain a lot of public attention. If anything, it might provide an argument for more expeditious transition to biometrics. Even, attempted 3-D printing of an ATM skimming machine, does not seem like something one would particularly worry about.

Some cases, however, might cause a valid concern. The consumer-oriented, Thing-O-Matic printer from Makerbot, although not as sophisticated as its commercial counterparts, is capable of printing firearms. A group called Defense Distributed published the schematics for a printable plastic gun. It is not hard to foresee that as the technology improves, it could make metal detectors useless and nullify many of the gun-control measures debated by lawmakers (Rayner 2013).

Similarly, a user of a site dedicated to sharing 3-D models for others to print out at home, posted plans for printing a magazine for an AR-15 rifle (Daw 2011). A fully automatic AR-15 can fire eight hundred bullets a minute. The posted model held five rounds of ammunition and thus was entirely legal. However, the file could be easily modified to hold fifteen or even more rounds. The federal law stipulates that possessing any magazine containing more than ten rounds was illegal. Another user of the same website posted a model for printing a part called the lower receiver for the AR-15. This is quite significant because typically it is possible to purchase any part of the mention rifle, for example, at gun shows without any records except that lower receiver. By printing out the lower receiver of an AR-15, one could potentially build a fully functional and yet unregistered AR-15.

Obviously, sharing questionable files and design blueprints that might be considered potentially dangerous is not anything new and certainly was not triggered by the emergence of digital manufacturing. It rather has to do with the notion of unrestricted Internet and as such has been tolerated. The practical impact of such exchange, however, was negligible simple because making them required often quite a high degree of technological sophistication. Things are, however, about to change dramatically our entire notion of controlled objects and substances will have to be revised. How are we to control or regulate something while we cannot control its distribution? After all, in a do-it-yourself environment, files are disseminated and modified and not the actual objects.

As with the issue of gun control, we ought to rethink or notion of controlled substances. Lee Cronin, at the University of Glasgow, has been experimenting with something he calls *reactionware*, which as he claims will allow people to print their own medication at home. A 3-D printer deposits a sequence of chemical agents into special gel chambers that create a controlled reaction. Cronin says that, eventually, consumers will be able to download a formula drug from a pharmaceutical company and print out the medication on their own. This surely holds great promise for patients, drug researchers and developing countries in need of medicines ("DIY drugstores in," 2012).

A variety of molecules have already been made, including some anti-cancer drugs. How does printing drugs actually work? Cronin's team used 3-D printer to manufacture the mention *reactionware*. These are essentially miniature containers with the chemicals that drive the reactions already built into them.

As Cronin points out, "by making the vessel itself part of the reaction process, the distinction between the reactor and the reaction becomes very hazy. It's a new way for chemists to think, and it gives us very specific control over reactions because we can continually refine the design of our vessels as required. For example, our initial reactionware designs allowed us to synthesize three previously unreported compounds and dictate the outcome of a fourth reaction solely by altering the chemical composition of the reactor ("DIY drugstores in," 2012)."

It is fair to say the technique is currently at an early stage of development. However, Cronin claims, since 3-D printers are becoming increasingly common, it's entirely possible that, in the future, we could see this chemical engineering trickling down to small commercial and individual users. Eventually 3-D printers could revolutionize access to health care in the developing world, allowing diagnosis and treatment to happen in a much more efficient and economical way than is possible now. The entire chain of drug manufacturing, marketing and delivery can be potentially disrupted, not to mention the system of patents and trademarks that is at the center of the current business model of the pharmaceutical industry. On the grass root level, domestic 3-D printers could turn into miniature medication factories, perhaps aided by some kind of software applications that would allow people to create the medication they need.

The prospect of personal pharmacy dispensing medication at home might seem very appealing. We must, however, consider the possibility of terrible abuses that might occur. How would we possibly monitor such process? Are we to part with the notion of controlling substances altogether?

It is crucial to recognize that most potential legislation might fail to keep pace with the rapidly expanding applications of digital manufacturing. As far as the misuse of drugs is concerned, some proposals include programming printers that specialize in medication to self-regulate against misuse.

Hod Lipson of Cornell University suggests that it would be more effective to introduce security regime that tracks gunpowder rather than firearms. As chemical-sensing technologies improve, he argues, detecting gunpowder may prove to be a more realistic security measure than hoping to outlaw continuous transformations of gun designs (Editors 2013). It is not clear, however, how tracking and regulating the movement of gunpowder on the marker would be less problematic than tracking firearms manufacturing. Apparently the assumption is that gunpowder is easier to detect at screening posts than a plastic gun.

Jonathan Zittrain of Harvard University offers a more compelling suggestion. He proposes that all 3-D printers should require connection to the Internet in order to function. Then, the printers should be programmed to check with a list of prohibited items online before printing a new object. This sounds promising, however, one must keep in mind that such solution would require building and maintaining a massive data base and it could be vulnerable to potential abuses as well (Hoffman 2013).

As the new technology makes crossing the boundaries of what permissible easier than before, how ought we to proceed on this shaky grounds? Should we introduce more regulations and how we can enforce it or rather try to outsmart the potential abusers? The problem is that digital manufacturing is still in its nascent stage and we are not fully aware of its potential capacities.

Today, we can only see the tip of the iceberg of what 3D printing can do. It would have been nearly impossible to predict the modern Internet from the barely connected world of the early 1990s. Similarly, we might think it is too soon for us to predict with any accuracy the real uses-and abuses-of a world where every home has a 3-D printer. If we move prematurely to regulate 3-D printers, we could fail to protect against the real dangers of the technology and at the same time limit the potential of 3-D printing to change the world for the better. We shouldn't get too anxious about something before we even fully understand what it is. Thus, to over-legislate now, aside from problems with the enforcement, would potentially mean to impede the process from which we could enormously benefit.

For now at least, the prudent way to proceed is to monitor the process very carefully and respond to challenges posed by 3-D printing as they arise with creativity and flexibility. Focusing on strategies to out-innovate the malefactors is possibly more feasible and beneficial way than legislating solutions for potential dangers— real and imagined.

7   Can 3-D Printing Change Our Perspective on Intellectual Property?

In the current patent system, there are major obstacles individuals with innovating ideas face in realizing their projects. For someone who wants to improve existing design, the proliferation of patents makes the process of obtaining permissions and licenses painful and costly. Some potential innovators may be discouraged by this daunting task and actually abandon their projects at the stage of a blueprint.

The existing patent law was conceived with the assumption that, in order to have technological progress, individuals or companies have to be rewarded for the disclosure of their inventions by granting them exclusive rights for some periods of time. Although it seems dubious that the creativity would cease in the absence of financial incentives, it certainly wasn't unreasonable to assume that inventors need to be rewarded for investments made while designing and prototyping.

How does 3-D printing challenge this model? Let's consider the issue of copying. With the help of 3-D scanners, it is relatively easy to translate the precise dimensions of an existing object into a digital file and upload it for sale. Then anyone could download the file and print an exact copy of things that are often subject to copyrights. One could print out a duplicate an expensive designer gadget just for personal use or sell a number of them at a steep discount. Obviously, counterfeiting is nothing new and different business models as well as governments have been dealing with it more or less successfully. What will change about it, however, as 3-D printing becomes common, is the scale of the phenomenon and the parties involved. Today, those that engage in counterfeiting are essentially replicating the established business model. After all, the counterfeiting enterprise requires substantial investments, connections and know-how. 3-D printing makes it much easier for anyone to create physical objects and the current system might very likely fail to sustain the pressures of the approaching explosion of innovation.

But can anyone possibly stop do-it-yourself copying, where digital files are copied, shared and modifies, whereas the printed product is not actually exchanged? Some think that we cannot stop it and we should not anyway. Michael Weinberg from Public Knowledge thinks that rights-holders' current control may be more illusory than it appears. Recognizing that fact early could be a competitive advantage. The "long tail" of things can't be easily controlled or monitored—he claims. In some cases striking a deal with those that infringe on a patent may prove to be more effective that litigation. Don't fight it, Weinberg suggests, simply embrace it: "One way to do this would be to create a transparent, available license. Simply put, HBO could make an offer to any designer: turn Game of Thrones into a product. Let us know about it. Sell it. And give us a cut. They could even include a caveat excluding some categories of products and reserving the right to revoke the license for specific products they find problematic. Even with those limitations, such an offer could lead to an explosion of creative objects. HBO would benefit from the increased publicity from the objects and, not incidentally, from a percentage of the sales. Designers would benefit from the knowledge that they could actually bring their dream product to market at the beginning of the design process. HBO and any individual designer would be free to negotiate a different arrangement, but the offer would create some certainty without requiring an additional negotiation (Weinberg 2013)."

Not everyone is enthusiastic, however, about such creative anarchy, especially that there would be virtually no way besides litigation to persuade individuals to make such arrangements in case they are not willing to. Instead, some advocate for ways to adapt the patent system to the changing environment and make it less stringent. More flexible utility models, already adopted in some countries, could serve as a starting point. Like patents, utility models, grant the holders the exclusive right to produce, use and market their inventions. But there are a few crucial differences. The rights are cheaper to obtain and their term is shorter -typically seven to ten years, as opposed to twenty years in patent system; they are also not renewable. In addition, they are easier to obtain since the novelty aspect of the invention is not scrutinized as thoroughly. Anyone, however, can challenge whether the registered invention actually does meet the standards of novelty, non-obviousness and utility. The losing party is responsible for paying for the proceedings.

The question is not whether 3-D printing will change how we manage intellectual property rights, but rather how do we respond to this change? It seems imperative that future legislation that will have to address the changing economic model strikes a delicate balance between preserving some form of intellectual property rights and expanding the freedom of the common creativity. Leaning towards any of the extremes might be detrimental to progress. Make the patent law to stringent and innovation might suffocate; allow creative anarchy and many project might be abandoned without the prospect of investment returns or financial incentives. Perhaps the industry should consider producing 3-D printers that require connection to the Internet in order to function as well as programming them to check with a global IP registry to make sure the user is not infringing on anyone's rights before modifying and printing a new object. Obviously, the current IPR would have to evolve as well. The rights should be granted for significantly shorter time than currently, perhaps ten years, after which period the design would become a public domain. Building and maintaining a large IP registry would undoubtedly pose some challenges. However, all parties would benefit from the system, particularly if the licenses were granted automatically and priced sufficiently low. If the system worked somewhat like iTunes music store, we can be moderately confident that common creativity would thrive while there financial incentives for inventions are preserved.

8 Navigating the Changing Landscape

With additive technology and digital design we are quickly changing the manufacturing landscape. Soon this technology will alter our lives. Learning how to navigate this new territory might be both captivating and frightening. Rapid prototyping is accruing across vast range of industries, significantly reducing the cost research and development.

It becomes clear that the entire business model will be revolutionized dramatically. Printing on demand might eventually eliminate assembly lines and radically shorten supply chains. In the process countless jobs associated with traditional manufacturing will be eliminated. In time, however, as with most technological breakthroughs, the transformation might create new jobs in new industries.

With the great promises of 3-D printing come ethical and legal concerns. The intellectual property law will face plenty of new challenges and most likely will have to adapt in order to survive. Entire new categories of products will be newly subject to counterfeiting and customer safety might be compromised due to massive customization. The rules of liability will become fuzzy as digital designs will be shared across national borders, modifies and printed with possibly varying quality.

The medical uses of 3-D printing are both awe-inspiring and petrifying. Printers are already being used to manufacture various prosthetics and doctors and scientists are experimenting with bio-printing human tissue and organs (see Fig. 9.6). This holds great promise of alleviating many of the current pathologies associates with transplants.

Digital manufacturing surely has not revealed its full potential yet. It is difficult to imagine all the creative applications future users of 3-D printers will come up with. *Fabricated*, a not-fiction book by Lipson and Kurman that reads like science-fiction, invites us to entertain the vision of the world reinvented by 3-D printing:

digital cuisine, robots that walk out of the printer, and responsive smart materials (Lipson and Kurman 2013). As far-fetched as it sounds, it might be just the tip of the iceberg of what the prospect of this technology holds. With Google Glasses and apps turning pictures into printable files, school children might be able to make replicas of artifacts seen on a museum class trip. All that the future inventors and manufactures might need to create sophisticated designs is software that can read their intentions directly from their brainwave. This could empower future successors of Da Vinci to unleash their imagination and creativity. This revolutionary technology will usher us into the new and exciting age of creative commons. Like anything else, it is up to us to decide how this technology will be used for better humanity or destruction.

References

Aamoth, D. (2013, June 27). Buttercup the duck gets a 3-d printed foot. *CNN Tech*, Retrieved from http://www.cnn.com/2013/06/27/tech/innovation/buttercup-duck-3d-foot/index.html.

Anderson, C. (2012). *Makers: The new industrial revolution*. New York: Crown Business.

Atala, A. (2012). *Printing human kidney* [Web]. Retrieved from http://www.bbc.com/future/story/20120621-printing-a-human-kidney.

Barnatt, C. (2013). *3D printing: The next industrial revolution*. ExplainingTheFuture.com.

Crook, J. (n.d.). Retrieved from http://techcrunch.com/2013/01/20/the-worlds-first-3d-printed-building-will-arrive-in-2014-and-it-looks-awesome/.

Daw, D. (2011, October 10). Criminals find new uses for 3d printing. *PCWorld*, Retrieved from http://www.pcworld.com/article/241605/criminals_find_new_uses_for_3d_printing.html.

DIY drugstores in development at the University of Glasgow. (2012, April 16). Retrieved from http://www.gla.ac.uk/news/headline_230503_en.html.

Editors. (2013, May 13). That 3-d printed gun? it's just the start. *Bloomberg*, Retrieved from http://www.bloomberg.com/news/2013-05-13/that-3-d-printed-gun-it-s-just-the-start.html.

Excell, J. & Nathan, S. (2010, May 24). The rise of additive manufacturing. *The Engineer*, Retrieved from http://www.theengineer.co.uk/in-depth/the-big-story/the-rise-of-additive-manufacturing/1002560.article.

Greenberg, A. (2012, July 7). Hacker opens high security handcuffs with 3d-printed and laser-cut keys. *Forbes*, Retrieved from http://www.forbes.com/sites/andygreenberg/2012/07/16/hacker-opens-high-security-handcuffs-with-3d-printed-and-laser-cut-keys/.

Hoffman, R. (2013, February 26). [Web log message]. Retrieved from http://www.bna.com/blogs_post.aspx?id=17179872508&blogid=12884902340.

Lipson, H., & Kurman, M. (2013). *Fabricated: The new world of 3d printing*. Indianapolis: John Wiley & Sons.

Marchione, M. (2013, May 22). *Doctors save life of Kaiba Gionfriddo, Ohio boy, by 3-D 'printing' him an airway tube*. Retrieved from http://www.huffingtonpost.com/2013/05/22/3d-print-airway-tube-kaiba-gionfriddo_n_3322217.html.

Rayner, A. (2013, May 6). [Web log message]. Retrieved from http://www.theguardian.com/world/shortcuts/2013/may/06/3d-printable-guns-cody-wilson.

Rifkin, J. (1995). *The end of work: The decline of the global labor force and the dawn of the post-market era*. Putnam Publishing Group.

Rifkin, J. (2013). The third industrial revolution: How the internet, green electricity, and 3-d printing are ushering in a sustainable era of distributed capitalism. *The World Financial Review*, Retrieved from http://www.worldfinancialreview.com/?p=1547.

Ruz, C. (2013, June 14). 3D printing powered by thought.*BBC*, Retrieved from http://www.bbc.com/future/story/20130613-3d-printing-your-thoughts.

Transplant jaw made by 3d printer claimed as first. (2012, February 6). Retrieved from http://www.bbc.co.uk/news/technology-16907104.

U.S. Patent 4,575,330 ("Apparatus for Production of Three-Dimensional Objects by Stereolithography").

Weinberg, M. (2013, March 26). [Web log message]. Retrieved from http://publicknowledge.org/blog/turning-3d-printed-copyright-infringers-partn.

Chapter 10
The Shape of the Sound
of the Shape of the Sound …

Stephen Barrass

1 Introduction

Digital fabrication is typically considered a one-way process, from the digital to the physical object. But could the process be considered as a transition between different states of the same artifact? The difficulty is that the 3D structure of a physical object is static, frozen in time. It cannot morph in response to changes in parameters like a digital structure can. However there is an aspect of every physical object that is temporal and dynamic —the sounds it makes. Physical acoustics are influenced by shape, size, material, density, surface texture and other properties of an object. Larger objects produce lower pitched sounds, metal objects are louder than plastic, and hollow objects produce ringing sounds. The acoustic properties of an object may be analyzed with spectrograms and other signal processing techniques. A spectrum contains all the information required to re-synthesize the sound from simple sine tones, and this is the theoretical basis for electronic music synthesizers. Could the spectrum recorded from a sounding object also contain the information to reconstruct the object that made the sound? This speculation leads to the idea to digitally fabricate an object from a sound recording. A sound could then be recorded from the new object. What would happen if another object was then fabricated from that sound? This recursive process of digital fabrication would generate an interleaved series of shapes and sounds shown in Fig. 10.1.

The rest of this paper describes experiments that explore this idea. The background section describes related concepts of synaesthetic transformation in painting, music and sculpture. It also describes previous work on sculptural 3D representations of music, and the digital fabrication of acoustic phenomena. The next section describes a first experiment to digitally fabricate a bell. This is followed by an experiment that develops a recursive method for generating a series of bells in which each bell is shaped by the sound of the previous bell in the series. The process is broken down into stages with parameters that can be adjusted to explore the space

S. Barrass (✉)
University of Canberra, Bruce, Australia
e-mail: stephen.barrass@canberra.edu.au

N. Lee (ed.), *Digital Da Vinci*, DOI 10.1007/978-1-4939-0965-0_10,
© Springer Science+Business Media New York 2014

Fig. 10.1 An interleaved recursive series of shapes and sounds

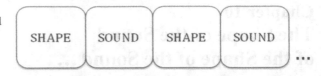

Fig. 10.2 Rhythmic Composition in Yellow Green Minor – de Maistre (1919)

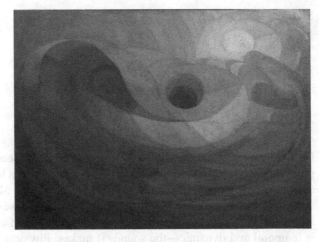

of possible outcomes. The discussion reflects on the results of the experiments, identifying theoretical issues and directions for further research.

2 Background

Wassily Kandinsky's invention of abstract painting was inspired by the abstract structure of music, and in his writing he refers to the synaesthetic composer Alexander Scriabin's 1915 score for "Prometheus: a Poem of Fire" which included a color organ that projected arcs and waves of color onto overhead screen in time to the music. The first abstract paintings in Australia were also inspired by music. Roy de Maistre's painting "Rhythmic Composition in Yellow Green Minor" featured in a controversial exhibition in Sydney in 1919 (Edwards 2011). His interest in relations between sound and color may have been inspired in part by his attendance one year beforehand at recitals on the color organ by Alexander Hector in 1918. De Maistre developed a formal Color Sound theory in studies such as Rainbow Scale D# minor—F# minor, and his works were popularly known as "paintings you could whistle" (see Fig. 10.2). Some of his other musical paintings include "Arrested Phrase from a Haydn Trio in Orange-Red Major", "Colour Composition Derived from Three Bars of Music in the Key of Green", and "The Boat Sheds, in Violet Red Key".

Fig. 10.3 Cthuga-Burfitt (1993)

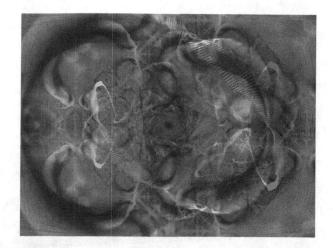

In 1993 the Australian coder Kevin Burfitt released the open source music visualization program Cthuga that was the forerunner of the visualization plugins in media players such as iTunes, Windows Media Player and VLC today (Burfitt 1993). Music visualizations map the loudness and frequency spectrum of sound into 3D graphics and image effects. The peer competition within the Cthuga community, and the ongoing commercial competition between large companies has resulted in high production values and well developed aesthetics in music visualizations (see Fig. 10.3).

Computer programs have also been used in the inverse transformation from graphics into sounds. The UPIC program, developed by algorithmic composer Iannis Xenakis in 1977, allowed waveforms and volume envelopes to be drawn on a computer screen to be electronically synthesized. (Xenakis 1971) An example of Xenakis' graphic music is shown in Fig. 10.4.

The representation of sound in visual form is extended to three dimensions in the Sibelius Monument created by Finnish sculptor Eila Hiltunen in 1967 to capture the essence of the music of the composer Jean Sibelius. The unveiling of the sculpture, constructed from more than 600 hollow steel pipes welded together in a wave-like pattern, sparked debate about the merits of abstract art that resulted in the addition of an effigy of Sibelius (see Fig. 10.5).

Digital fabrication provides a new way to create physical objects from sound. A search for "sound" in the Shapeways.com community for digital fabrication returns a set of 3D models titled 12, 24 and 48 Hz (shown in Fig. 10.6.) constructed from images of vibrations on the surface of water (Rubenacker 2012).

A further search for "music" on Shapeways returns several flutes, pan-pipes and whistles that may be fabricated in either plastic or metal. There is also a wind-chime fabricated in glass or ceramic. These examples show the potential to use 3D CAD tools and personal fabrication services to custom design sonic objects and acoustic structures.

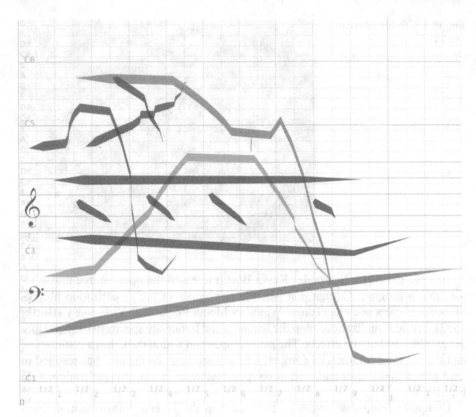

Fig. 10.4 Mycenae Alpha – Xenakis (1978)

Fig. 10.5 Sibelius Monument – Hiltunen (1967)

Fig. 10.6 48Hz - Rubenacker
(2012)

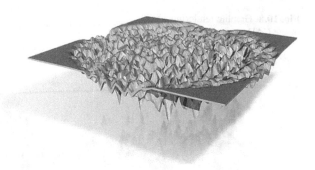

Fig. 10.7 Federation Bells –
McLachlan (2000)

Neil McLachlan used a CAD package and computer modeling to design a set of 200 harmonically tuned bells for the Federation Bells installation in Melbourne in 2000, shown in Fig. 10.7. He identified the geometric factors that influence the harmonics as wall thickness profile, wall curvature, conical angle, the circumference of the opening rim, the thickness of the rim, and the overall width and height of the bell (McLachlan 1997). Bells are complex 3D shapes that flex in 3 dimensions, and they are much more difficult to tune than one-dimensional wind or string instruments. Tuning a bell was traditionally done by skilled craftsmen who manually lathed the thickness profile of a cast bell. Due to the high costs of casting bells in the modern era, McLachlan manufactured CAD bells by pressing sheet metal, which had the advantage of very consistent geometry. The fixed thickness required tuning of harmonics by shaping the wall curvature, rather than lathing the thickness (McLachlan 2004).

Advances in digital fabrication technology have brought new materials, such as stainless steel, bronze, silver, titanium, glass, and ceramics. The introduction of metal shaping technologies in the iron and bronze ages resulted in the invention of bells, gongs, singing bowls and other resonating musical instruments. Could the introduction of metals in digital fabrication herald a new era of sounding objects that could not be arrived at by manual crafting?

Fig. 10.8 Graphic rendering of the CAD mesh of Bell00

3 Digital Fabrication of a Bell

This section describes an experiment to extend previous work on CAD bells by digital fabrication, with a view to more complex sounding objects in the future.

Digital fabrication places constraints on size, thickness and level of detail, depending on the material. The Shapeways.com service constrains stainless steel to a maximum bounding box of 1000 × 450 × 250 mm, wall thickness of 3 mm, and detail of 0.6 mm. This is quite limiting but does allow for the fabrication of small bells.

A bell shaped 3D mesh was constructed from graphic primitives using the processing.org open source environment for graphic programming. The outer hemispherical shell with diameter 42 mm and height 34 mm was duplicated, scaled and translated to make an inner shell. The rims of the outer and inner shells were "stitched" together to make a watertight shape. A handle was added so the bell could be held without being damped. The digitally constructed bell, shown in Fig. 10.8, was saved as a CAD file in STL format.

The CAD file is limited to 64 MB and the polygon count to less than 1,000,000 for uploads to the Shapeways site. The high resolution mesh was reduced in size and count by merging close vertices in the Meshlab open source system for editing unstructured 3D meshes (http://meshlab.sourceforge.net/). The mesh was then checked to be watertight and manifold using the Netfabb software for editing and repairing 3D meshes for additive manufacturing (http://www.netfabb.com/). This carefully prepared CAD file was then uploaded to Shapeways, and fabricated in stainless steel with bronze colouring, to produce the first prototype of a digitally fabricated bell shown in Fig. 10.9.

When the bell was tapped with a metal rod it produced a ringing tone. The sound was recorded at 48 kHz sampling rate with a Zoom H2 recorder in a damped room. The recorded waveform in Fig. 10.10 shows that it rings for about 1 s.

The spectrogram, in Fig. 10.11, shows partials at 2.971, 7.235, 13.156 and 20.356 kHz. The first rings for ~1.2 s, second ~0.75 s, third ~0.5 s and fourth ~0.2 s. The temporal development of these partials produces the timbral "color" of the bell. Although the partials are not harmonic, the bell does produce a clearly pitched tone.

Fig. 10.9 Digitally fabricated bell

The Long Term Average Spectrum (LTAS) is a 1D summary of the spectrogram. The LTAS in Fig. 10.12 shows the peak amplitude for the four main partials, along with the four main regions of resonance that produce the ringing timbre of the bell.

The prototype demonstrates that a bell can be digitally fabricated, and opens the door to more complex acoustic objects that cannot be manufactured or made manually.

4   Recursive Bells

This section presents an experiment to design a recursive series of bells where each bell is shaped by the sound of the previous bell in the series.

The stages of the recursive process are shown in Fig. 10.13. The process begins with the CAD file specifying an initial bell, labeled as BELL 0. The CAD file is fabricated as a physical shape, SHAPE 0, which is the stainless steel prototype bell constructed in the previous section. The sound of SHAPE 0 is generated by tapping the bell, and recording it as SOUND 0. This sound is then transformed into PROFILE 1 by a process labeled XFORM. Then PROFILE 1 is added to BELL 0 and the new CAD file is fabricated as SHAPE 1, which is the next bell in the series. SOUND 1 is then recorded by tapping SHAPE 1, and XFORMed to create PRO-FILE 2, which is added to BELL 0 to create the second recursive bell. This recursive process can be repeated ad. infinitum to produce a series of interleaved SHAPES and SOUNDS generated from each other.

4.1   XFORM

The XFORM is a mapping from sound into a thickness profile that can be added to a bell shape to change the sound it makes.

The LTAS analysis of the prototype bell captures timbral features in a 1 dimensional format that can be used to algorithmically construct a thickness profile as

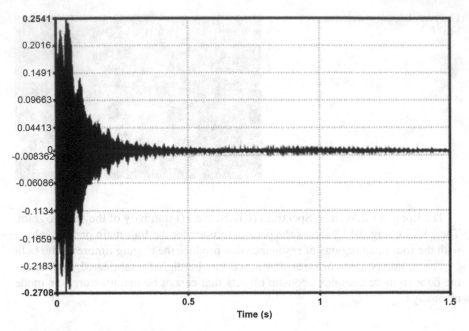

Fig. 10.10 Waveform of bell 0

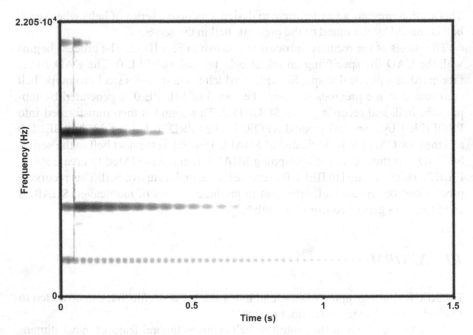

Fig. 10.11 Spectrogram of bell 0

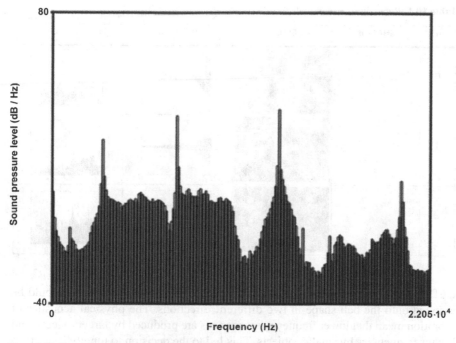

Fig. 10.12 Long Term Average Spectrum (LTAS) of bell 0

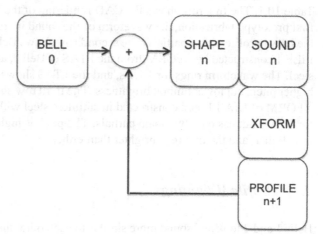

Fig. 10.13 Recursive fabrication process

Table 10.1 Recursive series of bells 0, 1, 2

n	SHAPE n	BELL n	SOUND n	LTAS PROFILE n+1
0				
1				
2				

a 3D mesh. The LTAS has low frequency and high frequency ends that could be mapped onto the bell shape in two different directions. The physical acoustics of vibration mean that lower frequency resonances are produced by larger objects, and higher frequencies by smaller objects. This led to the decision to tonotopically map the low frequency end of the LTAS to the large circumference at the opening rim, and the high frequency end to the smaller circumferences towards the crown.

The first experimental series of bells generated using this XFORM is shown in Table 10.1. The first row shows the CAD rendering of the basic bell, a photo of the first prototype fabrication, the waveform of the sound it produces when tapped, and the LTAS profile with 4 partials. The second row shows Bell 1, with thickness PRO-FILE 1 constructed by XFORM from the LTAS of Bell 0, and fabricated in stainless steel. The waveform rings for ~1.5 s, and the LTAS shows 3 partials that produce a higher pitch, but lower timbral brightness. The third row shows Bell 2 shaped by the XFORM of LTAS 1, and constructed in stainless steel with gold color. Bell 2 rings for 0.75 s, but has only two main partials. The pitch is higher than Bell 0 and lower than Bell 1, and the timbre is brighter than either.

4.2 Profile Weighting

Bells 1 and 2 look and sound more similar to each other than expected. The weighting of the shape profile relative to the bell template can be adjusted in the mesh generating program. The ability to alter this weighting has been added to the process diagram as a parameter labeled T in Fig. 10.14.

The next experiment tested the effect of varying parameter T on the sound of Bell 2. An alternative Bell 2+ was fabricated with T double the previous level, thereby doubling the geometric effect of the PROFILE generated from the sound of Bell 1.

Fig. 10.14 Process with profile weighting T

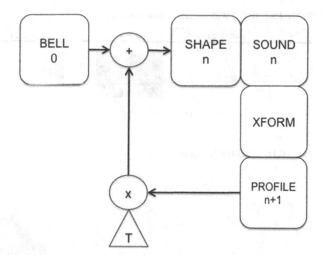

The results in Table 10.2, show an amplitude modulation in the ringing sound that is heard as a tremolo effect. There has also been an increase in the frequency of the two main partials. Bell 2+ is distinctly different in timbre from Bell 2, and Bell 1.

This result suggests that increasing T may generate more variation in the series of shapes and sounds. To explore this further the value of T was raised to 3x and used to generate the next bell in the series. The CAD rendering of Bell 3++, shown in Fig. 10.15. has wide flanges that indicate that raising T too high could transform the geometry beyond the point where it will function as a bell. On the other hand, these flanges may introduce unusual timbral effects, such as tremolos and vibratos, that are not heard in conventional bells. This bell was too thin to 3D print. Recursive processes that are theoretically infinite can be terminated by practical limits such as the memory in a computer, or, in this case, the physical limits of the fabrication process and material.

4.3   Material

The Bells in the experiments have so far been fabricated in stainless steel. However, other materials, such as ceramic and glass, also have good acoustic properties. The recursive generation process is updated with a stage for materials in Fig. 10.16. What is the effect of using these materials on the acoustics of the bell?

Bell 2 was re-fabricated in ceramic. This version of the bell is smoother and has less detail, as can be seen in Table 10.3.

Tapping the ceramic Bell 2 produced a short, sharp, high pitched, percussive sound very different from the ringing produced by the stainless steel version. The LTAS profile has 3 partials that look generally similar to previous bells. However the short duration makes it difficult to hear spectral details. The reduced detail of the ceramic fabrication effectively low pass filters the LTAS profile. Does this reduced

Table 10.2 Bell 2 with doubled parameter T

n	SHAPE 2+	BELL 2+	SOUND 2+	LTAS PROFILE
2				

Fig. 10.15 CAD rendering of
Bell 3++

detail have a perceptible effect on the sound the bell makes? This could be answered
by fabricating a low-pass filtered version of Bell 2 in metal, and then comparing the
sounds produced by the smoothed and original bells.

5 Discussion

The effect of varying the T parameter raises the question of whether the series will
converge to an attractor shape, traverse a contour of endless variation, or diverge to
a point of destruction? Is there a value of T on the boundary between convergence
and divergence? Is the recursive process a random walk or does it have a trajectory
of some kind?

If the series does converge, the bell will produce a sound that has an LTAS profile that is identical to its own thickness profile. The shape of this bell is a blueprint
for the sound it produces, and the sound contains the blueprint for the bell that
produced it. This attractor bell and its sound would be bilateral transformations of
the same trans-phenomenal object. Does such an object actually exist, and can it be
found with this process?

The XFORM mapping between the sound and shape in these experiments has
been a simple mapping of LTAS to thickness profile. The decision to map the LTAS
in one direction raises the question of whether mapping it in the opposite direction
would make a difference. There are also other ways that features of a recorded
sound could modify the acoustics of a bell. The audio waveform could be wrapped
in a spiral down the bell shape, etching into the profile in a manner similar to a

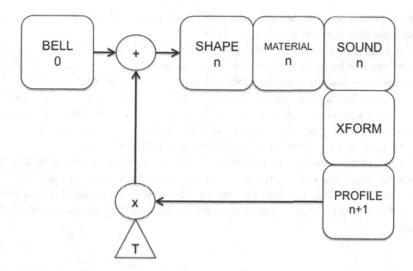

Fig. 10.16. Recursive process incorporating material

Table 10.3 Bell 02 fabricated in ceramic

n	SHAPE 2	BELL 2 ceramic	SOUND 2 ceramic	PROFILE 3 ceramic
2				

needle groove on a wax cylinder or record. The frequency axis of the 2D spectro-gram could be assigned to the radial angles of the bell with the amplitude affecting the profile in the radial directions. Other kinds of timbral analysis could be used, such as mel frequency cepstral co-efficients (MFCC), or centroid, flux, kurtosis, and skew.

6   Conclusion

These experiments to generate a recursive series of bells and sounds have identified generic stages in a systematic process. The XFORM stage is a mapping between sound and shape. The T parameter controls the level of feedback in the recursive circuit, and the amount of variation in the shapes and sounds that are generated. This parameter may also affect whether the series converges, traverses a contour of variation, or diverges to destruction. The material has a significant effect on the acoustics of the object, and different materials may cause convergence to particular attractor nodes, for example the lack of detail in ceramic shapes and sounds may cause rapid degeneration to a singular point.

The bells in these experiments open the door to the design of more complex shapes than can be made with conventional manufacturing techniques. The geometry of acoustic shapes could be generated using a 3D fractal such as the Mandelbulb, or a rule based L system. These shapes can have complexity that is beyond the state of the art in acoustic simulation with finite element meshes. Digital fabrication allows rapid prototyping of physical objects that could allow research on the acoustics of shapes that are more complex than hitherto been possible.

These experiments have raised many theoretical questions to guide further experiments which are still in progress. Can the recursive process be used to find a trans-phenomenal artifact where the acoustic response contains the blueprint of the object that produced it? What new shapes and sounds will be generated through this process?

References

Burfitt, K. Cthuga, Retrieved 21 May 2014, http://www.afn.org/~cthugha.
Edwards, D. *Colour in Art: Revisiting 1919*. Art and Australia II: European Preludes and Parrallels, Diploma Lecture Series, Art Gallery of NSW, 2011.
McLachlan, N. *Finite Element Analysis and Gong Acoustics*. Acoustics Australia, 25, 3, 103–107, 1997.
McLachlan, N. *The Design of Bells with Harmonic Application of New Analyses and Design Methods to Musical Bells*. in Proceedings of the 75th Conference of the Acoustic Society of America, New York, 1–8, 2004.
Rubenacker, M. Retrieved 21 May 2014, http://www.shapeways.com/model/91488/48-hz.html
Xenakis, I. Musique. Architecture, rev. ed. Paris: Casterman, 1976 (first published in 1971). Published in English as Music. Architecture, tr. Sharon Kanach.Stuyvesant: Pendragon Press.

Chapter 11
Human-Robot Interaction in Prepared Environments: Introducing an Element of Surprise by Reassigning Identities in Familiar Objects

Mari Velonaki

1 Preamble

My fascination with projected and kinetic characters started in 1996. It all began with 'Red Armchair 4', an interactive installation that utilised speech recognition. In this work (Velonaki 2006) audience expectations were manipulated by withholding the full appearance/identity of the projected character, whose face was never revealed. In *Red Armchair 4*, the visitor walks into a red-lit room and is presented with a projected image of a woman in a black dress. She is viewed from the back, seated on a red armchair which ensconces her in a shell-like embrace. An identical chair is placed in the middle of the room, where the visitor can sit facing the back of the projected woman. On the floor there is a light box printed with the command words:

DECAYCONSUMETALKDANCESHRINKAWAKELOOKDIE

A microphone adjacent to the visitor's chair enables him/her to talk to the woman on the screen by forming sentences that include the command words chosen at random (for example: 'Talk to me' or 'Die for me'). Every time a participant says a command, the character is activated, moving from a still frame to a moving image. Her physical responses, however, are completely unrelated to participant's spoken requests. She runs her fingers through her hair, falls asleep on the arm of the chair, swings her legs. She appears to be in her own space, comfortably consumed by self-absorption. The only time she stands up and walks towards the participant, her head is out of the frame; the participant's desire to see her face is never satisfied.

To my surprise, I realised that although the projected woman's relationship to her visitors is unrewarding—they don't even see her face—they were fascinated by her 'cinematic' presence, and spent significant time in the gallery installation space, attempting to either control or 'communicate' with her. The spectator's physical and intellectual engagement became one of the main concerns of my research

M. Velonaki (✉)
University of New South Wales, Kensington, Australia
e-mail: mari.velonaki@unsw.edu.au

N. Lee (ed.), *Digital Da Vinci*, DOI 10.1007/978-1-4939-0965-0_11,
© Springer Science+Business Media New York 2014

and creative practice. His/her actions complete the work; the way s/he approaches the work, his/her physical placement in relation to the work, how much time s/he spends in the space, whether s/he engages with the work in a relaxed or tense manner. The spectator's behaviour also changes depending on whether s/he visits alone, in the company of friends, or surrounded by strangers. Even if the spectator abstains from physical engagement with a situated interactive 'character', choosing to stand in front of it and just look at it with interest, anticipation, disbelief or disapproval, still constitutes an interaction. The Greek word for interactive is αμφίδρομος, or amphi-dromos (amphi: around on both sides of, dromos: street or road) (Hoad 1986a; Magazis 1995). Thus it is defined as a middle point where two roads meet; this is the closest description of what the author's work tries to achieve. In English, the preposition 'inter' means 'between' or 'among' (Hoad 1986b). Inter-action, therefore, signifies between or among actions. A meeting point beyond action and reaction and prior to discourse, a brief moment of recognition between two parties.

The forms of my robotic works vary widely from wooden cubes to wheelchairs to porcelain-like statues. I never intended their appearance to realistically resemble a human, yet their behaviour has been strongly influenced by human patterns; although my robots don't appear realistically anthropomorphic (e.g. as an android robot), the design of their behaviour is consciously anthropocentric, believing that a participant will respond more immediately to a behaviour that can be assigned to a person. I describe three of my projects (2004–2013) in the following sections.

2 Fish-Bird (2004–2006)

With *Fish-Bird* (2004–2006) my work moved towards a different direction, from projected characters to three-dimensional kinetic objects that represent characters. In 2003, I started working at the Australian Centre for Field Robotics at the University of Sydney, forming a collaborative team with David Rye, Steve Scheding and Stefan Williams, roboticists at the Centre. Although we were from different disciplines, myself from media arts and interface design and my colleagues from robotics and automation, our shared goal was to better understand the complex space of human-robot interaction, and what elements could assist in triggering an engaging interaction.

Fish-Bird is an interactive autokinetic artwork that investigates the dialogical possibilities between two robots, in the form of wheelchairs, that can communicate with each other and with their audience through the modalities of movement and written text (see Fig. 11.1). The chairs write intimate letters on slips of paper that they then drop to the floor, impersonating two characters (Fish and Bird) who fall in love but cannot be together due to 'technical difficulties'.

Spectators entering the installation space disturb the intimacy of the two objects, yet create the strong potential for other dialogues to exist. The spectator can see the traces of their previous written exchanges on the floor, and may become aware of the disturbance that they have caused. Dialogue occurs kinetically through the

Fig. 11.1 Mari Velonaki, *Fish-Bird: Circle B—Movement C* (2004–2006). Interactive installation with two autonomous robots and distributed data fusion system

wheelchair's perception of the body language of the audience, and on the audiences reaction to the unexpected disturbance would be to converse about trivial subjects, like the weather... Through emerging dialogue, the wheelchairs may become more "comfortable" with their observers, and start to reveal intimacies on the floor again.

Each wheelchair writes in a distinctive cursive font that reflects its 'personality'. The written messages are subdivided into two categories: personal messages communicated between the two robots, and messages written by a robot to a human participant. The messages are an amalgamation of words, verses and sentences selected from a large database containing excerpts of the poetry of Anna Akhmatova, fragments of love-letters donated to the project by people over the period of three years, and text composed by me.

At the time, by choosing the wheelchair as the form for the robots, I aimed to introduce a new aesthetic proposition in robotics: one that was far removed from humanoid, android or pet-like robots. A wheelchair is the ultimate kinetic object, since it self-subverts its role as a static object by having wheels. At the same time, a wheelchair is an object that suggests interaction—movement of the wheelchair needs either the effort of the person who sits in it, or of the one who assists by pushing it. A wheelchair inevitably suggests the presence or the absence of a person. Furthermore, the wheelchair was chosen because of its relationship to the human—it is designed to almost perfectly frame and support the human body, to assist its user to achieve physical tasks that they may otherwise be unable to perform. In a similar manner, the Fish-Bird project utilises the wheelchairs as vehicles for communication between the two characters (Fish and Bird) and their visitors. One of my aims was to test the hypothesis that robot behaviour can be more important that appearance in determining levels of engagement in human-robot interaction.

The dialogical approach taken in this project both requires and fosters notions of trust and shared intimacy. It is intended that the technology used in the project will be largely transparent to the audience. Going further than a willing suspension of disbelief, a lack of audience perception of the underlying technological apparatus will focus attention on the poetics and aesthetics of the artwork, and will promote a deeper psychological and/or experimental involvement of the participant/viewer. Robots in the context of popular culture have historically been associated with anthropomorphic representations. Although they represent characters, the robots in Fish-Bird are not anthropomorphic, nor are they pet-like or 'cute'. The audience internalises the characters through observation of the words and movements that flow between the characters, and between the characters and the audience, in response to audience behaviour. Through movement and text the artwork creates the sense of a person, and allows an audience to experience that person through the perception of what is not present.

3   The Fragile Balances Series: Circle D and Circle E (2008–2010)

Fragile Balances was created as a companion work to Fish-Bird. I wanted to create two new embodiments of Fish and Bird that would act as avatars to enable the activation of their dialogues in locations remote from the robots. I also wanted to test agency in relation to physical appearance and, in particular, how people would respond to hand-held interactive objects. In Fish-Bird the participants were surprised in their encounter with what appeared to be regular wheelchairs when the wheelchairs started to move and respond to the participant's movement. They didn't expect the wheelchairs to move by themselves, follow them, avoid them by 'hiding' in corners, or print messages for them. I believe that the fact that the wheelchairs didn't look at all technological contributed to the element of surprise and subsequent engagement. In Fragile Balances I chose to design another object with a non-technological appearance, although it had to house highly-technological electronic modules. The choice was to work with wood, an organic traditional material.

Circle D is comprised of two luminous cube-like wooden objects that appear to be floating above the surface of a lacquered structure that perches on impossibly slender legs (see Fig. 11.2). Each object is comprised of four crystal screens where 'handwritten' text appears, wrapping around it conveying a playful sense of rhythm. The text represents personal messages that flow between the virtual characters of Fish and Bird, and in that sense each object is a physical embodiment of a character. The objects can be lifted from their wooden stand and handled freely by participants (see Fig. 11.3). Handling provides an interface that facilitates bidirectional communication between the participants and the artwork in a playful way.

Fig. 11.2 Mari Velonaki, *Circle D: Fragile Balances* (2008–2010). Interactive installation with two autonomous objects

If a gallery visitor picks up one of the cube-like objects from its floating base the text becomes disturbed and barely readable, influenced directly by the movement of the visitor's hands (see Fig. 11.4). The sensitive structure of the personalised messages flowing between the two fictional characters remains disturbed as long as the visitor moves or turns the object quickly or abruptly. The only way that the participant can allow the messages to again flow around the object is to handle it with care—gently and softly cradling the object in his/her hands in concert with the rhythm of the 'handwritten' messages. If visitors do not handle the luminous cube objects, the work stands on its own as a complete sculptural piece containing an internal kinetic element—the moving text. *Circle D* was never intended to be a 'gadget' or a game that gives rapid gratification; instead, the intention was to use the cube as an interface to slow people down, by creating an almost meditative space where pausing becomes rewarding.

In this installation the objects provide an interface that facilitates bidirectional communication between the participants and the autonomous objects. *Fragile Balances* deals with concepts of fragility, trust, and communication by playfully challenging the participants to pause and enter the rhythm of the floating words and the dialogues that they lead to.

Fig. 11.3 Mari Velonaki,
Circle D: Fragile Balances
(2008–2010). Interactive
installation with two autonomous objects

Fig. 11.4 Mari Velonaki,
Circle D: Fragile Balances
(2008–2010). Interactive
installation with two autonomous objects (detail)

Interaction occurs between artwork and audience through the reactive objects as information passes from the object to the participant and from the participant to object. Another linkage involves the latent relationship between the two participants—the objects become the medium for a participant to become aware of the existence of a virtual character—in this case, Fish or Bird. The moment that a participant chooses to pick up and hold one of the objects, s/he becomes an avatar for this character in the actual physical space of the installation. Should *both* participants then choose to vocalise their individually-received fragmented messages as they appear around the surface of their objects, or to move close to each other in order to read each other's messages, then the dialogue between the virtual characters is manifested and completed in the physical space. To reach into this fragile stream of text, the participant must attain a moment of stillness.

As in *Fish-Bird*, in *Fragile Balances* it was important that the technological apparatus was concealed, and therefore invisible to participants. This inspired adaptations in engineering and design to meet a set of aesthetic criteria on the physical manifestation of the two avatars. Each small cube conceals custom-built miniaturised microcomputers, accelerometer sensors, batteries, and circuitry for battery charging and power management. No external wires are visible—the design was purposely manipulated to eliminate the visibility of screws and other such traces of the assembly process, and the stand also functions as a concealed battery charger.

Circle E was created to provide an interface where participants could hand-write and 'post' their own messages to the Fish and Bird avatars of *Circle D* (see Fig. 11.5). *Circle E* is a wooden table-like object with a rotating brass drum partially sunk into it, a notepad and pencil are placed on its top and a 'postal bag' hangs under the object. Members of the audience are encouraged to write to Fish and Bird, or to their loved ones and donate their letters to the project by feeding them through the slot into the drum when it pauses momentarily (see Fig. 11.6). All the letters are scanned and, at a later stage, added as text to the dialogues between the Fish-Bird robot and the interactive cubes of *Circle D*.

Regardless of where the *Fragile Balances* series has been exhibited—Australia, Hong Kong and mainland China, Korea, New Zealand to date—it has been overwhelming to see not only that thousands of people have written messages, but the very personal nature of these messages, often accompanied by drawings, addressed to the Fish and Bird avatars that inhabit the cubes.

4 Diamandini (2009–2013)

With *Diamandini*, I wanted to make a new robot that would take the experimentation that began with *Fish-Bird* further by adding the element of interaction via touch (see Figs. 11.7 and 11.8). It was important that the interaction be one-to-one: one human, one robot.

Fig. 11.5 Mari Velonaki, *Circle E: Fragile Balances* (2008–2010). Interactive installation incorporating rotating brass drum

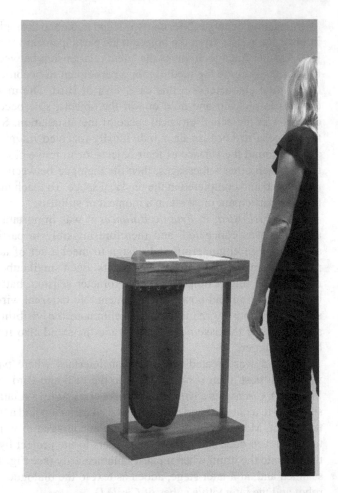

Fig. 11.6 Mari Velonaki, *Circle E: Fragile Balances* (2008–2010). Interactive installation incorporating rotating brass drum (detail)

Fig. 11.7 Mari Velonaki,
Diamandini (2009–2013).
Interactive humanoid robot.
(Image courtesy of the Victoria & Albert Museum)

The initial intent of the *Diamandini* project was to create a robot that was both non-representational and non-anthropomorphic. As I started experimenting with a variety of abstract sculptural forms, although interesting in shape and structure, I found it extremely difficult to assign behaviours to them that could lead to emotional activation of the spectator/participant to a degree that s/he would be touch and even embrace the robot.

These considerations influenced my decision to create a humanoid robot. This was a challenging decision, especially when deciding how the robot should look. I did not want *Diamandini* to have a typical humanoid robot aesthetic. After a long period of reflection I began to think of *Diamandini* as a female sculpture. In my mind *Diamandini* had a diachronic face that spans between centuries, a style that could be reminiscent of post-World War II fashion influences and, at the same time, with futuristic undertones. Most importantly, I didn't want *Diamandini* to look like a stereotypical humanoid robot. Her exterior is entirely made of a homogeneous

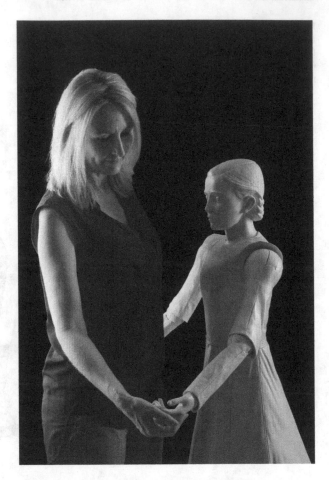

porcelain-like material that makes her look more like a floating figurine than a robot.

Diamandini is small–only 155 cm high. I wanted her figure to be small and slender so that people did not feel threatened by her when she 'floats' in the installation space. I wanted *Diamandini* to look youthful, but not like a child, and for her age not to be easily identifiable. In my mind *Diamandini* is between 20 to 35 years old.

The construction of *Diamandini* was a multi-stage process, involving a sculptured prototype terracotta head, a custom-tailored fabric dress made over a wooden armature, and high precision 3-dimensional laser scanning and manipulation of the scanned data, followed by computer-aided design modelling. The external shell was made using stereolithography–an additive manufacturing process that uses computer-controlled UV lasers to polymerise a resin in very thin layers. The exterior is treated with a porcelain-like material.

As it was important for Diamandini to appear to 'float' across the floor of an installation space and therefore a commercial motion base could not be used. An omnidirectional motion platform containing three computer-coordinated driven and steered wheels was designed and constructed. The omnidirectional motion base decouples Diamandini's facing direction and rotational motion from the direction and speed of her movement so that she can glide backwards, forwards or sideways, and transition smoothly between these movements.

Diamandini's arms were designed with four rotational degrees of freedom each–two at the shoulder and two at the elbow–so that the articulation of her arms was simple, yet able to make the range of gestures required for interaction. Sensing, including distributed touch sensing, and actuation will be added to the current prototype arms.

The project has also involved considerable research into touch sensing and the transmission and interpretation of social messages and emotions via touch. Techniques based on electrical impedance tomography (EIT) were developed that can be used to implement flexible and stretchable artificial 'sensitive skin' to facilitate the interpretation of touch by robots during human-robot interaction. A classifier based on the 'LogitBoost' algorithm was developed and used to identify social messages–such as 'acceptance' and 'rejection'–and emotions transmitted by touch to an arm covered by the EIT-based skin. Experiments demonstrated, for the first time, that emotions and social messages present in human touch could be identified with accuracies comparable to those of human-to-human touch (Silvera Tawil et al. 2012). Future work in the project will transfer these techniques to Diamandini.

To date Diamandini has been presented in only three contexts: as a prototype during ISEA 2011 in Istanbul, as an interactive sculpture in the Medieval and Renaissance Gallery at the Victoria & Albert Museum, and as a performative automaton at as part of *Time & Motion: Redefining Working Life* at the Foundation for Art and Creative Technology (FACT) in Liverpool. At the V&A Museum more than 2,500 people interacted with Diamandini during the "Digital Weekend". Her placement in a gallery, surrounded by medieval and renaissance sculptures, startled some people who at first mistook her for a statue. At FACT, Diamandini was gliding around the exhibition, pausing in front of artworks and then continuing her spatial exploration of the gallery in a choreographed manner, stopping only when her desired path was blocked by people.

5 Conclusion

I create technological objects that—at first glance—look familiar, and initially they may be perceived with a set identity. The element of surprise comes when people discover that their behaviours are different to what their appearance suggests. In situated interaction design this momentary element of surprise can significantly contribute in promoting an exploratory mode of interaction between a participant and an embodied agent. The most rewarding aspect for me is how this encounter produces unexpected behaviours from the human participants. What still fascinates me the most is how wonderfully unpredictable humans can be.

References

Hoad, T.F. (1986a). *The Concise Oxford Dictionary of English Etymology*, Oxford University Press, New York, 14.

Hoad, T.F. (1986b). *The Concise Oxford Dictionary of English Etymology*, Oxford University Press, New York, 239.

Magazis, George A. (1995). *Greek Word Master*, Efstathiadis Group, Athens, 279, 340.

D. Silvera Tawil, D. Rye & M. Velonaki. Interpretation of the modality of touch on an artificial arm covered with an EIT-based sensitive skin. *International Journal of Robotics Research*, vol. 31, no. 13, November 2012, pp. 1627–1642. D. Silvera Tawil. Artificial skin and the interpretation of touch in human-robot interaction. PhD Thesis, The University of Sydney, 2012.

Velonaki, M., 'Apples, wheelchairs and unrequited love'. In C. Hart (ed.), *Can We Fall in Love With a Machine?*, Pittsburgh Cultural Trust, Pittsburgh PA, pp. 85–90, 2006.

Chapter 12
The Messages of Media Machines: Man-Machine Symbiosis

Roman Danylak

1 Introduction

McLuhan's famous dictum, *the medium is the message* (McLuhan and McLuhan 1988), guides this chapter and its content. The statement, when paraphrased, may be understood as—it is the nature of a medium that decides what messages can be transmitted. This underlines the view that the *form* of the technology, its nature and characteristics, dictates the *content*. The focus here is to understand the computer as a medium—its form, supporting the greater ambition of defining its messages —its content. A key feature of content, it will be argued, is that personal identity is profoundly reflected by media content. In this way, man and media machines are inextricably linked.

A *medium*—that which is in between—refers to material and non-material processes for transmitting messages (Dictionary 2005) Media forms include text, film, television and radio. A *medium*, of which *media* is the plural, is defined as 'something in between'. A medium then, is that which carries a message between the maker/sender and the receiver of the information, enabling communication. The air that carries our verbal utterances is a medium; a clay tablet with impressions marking the number of head of cattle is a medium. Media are communication technologies with a long history and are used to create inventory and portability of information through the senses, primarily of what is seen and heard, but this now also includes touch. A medium simply means *that which is in between*, or that which is in the middle of two communicating individuals. In this way, paper is a medium carrying the message of written words to from the writer to the reader. Computers have a multimedia dimension in that many media, often interlinked, are present in the one machine (Dijk 2004).

Andersen (Andersen 2001) a computer semiotician, has made some salient observations about the computer as a medium. He has stated that:

> Computers are new media, and the human aesthetics of new media normally evolve out of older media by borrowing and restructuring older techniques.

R. Danylak (✉)
University of Technology, Sydney, Ultimo, Australia
e-mail: roman@emotional-computing.com

N. Lee (ed.), *Digital Da Vinci*, DOI 10.1007/978-1-4939-0965-0_12,
© Springer Science+Business Media New York 2014

However new media normally have their own unique artistic effects, and it takes time to discover what they are. (Andersen 2001, p. 2)

The approach adopted here is to examine 'older media' namely film, then later painting, to attempt to structure the unique expressive and design the effects of computer as a medium the final effect termed *interactive film*. The discussion to follow will be structured in three parts: first, a brief survey and examination of old media (400–500 years old) in particular, the book, a precursor of digital simulation and how its messages has brought us to see and experience the world as we do with particular reference to the notion of the *individual*; secondly, a discussion on the nature of painting mostly as a modern medium and how that medium and its message behave; thirdly, presenting a number of points on the message making characteristics of the medium of interactive film and what significance tis may have for the messages of identity.

2   The Message of the Printed Book: The Rise of the Individual

It is worthwhile examining the last great period of new media invention—what is now old media—and its effects upon society as a result of the messages that these media conveyed—towards building a model of medium and it message in computing. In European history, it was the invention of the mechanical printing press by Gutenberg in 1450 (McLuhan 1964) that caused a revolution in individual perception giving rise to what we now know as the *individual*. The book, previously a hand- made object and therefore rare, was restricted knowledge and poorly disseminated kept mostly in monastic cloisters. With the *advent* of mechanical metal cast moveable type, books become more readily available and by 1623 the publication of Shakespeare's first folio copies occurred, some of which still exist to this day (Shakespeare and Mouston 1995). This ready, privately available information gives rise to the individual because it is the individual who can think and interpret for themselves the message of the author now having access to an otherwise absent medium. This signals a departure from the rigid preceding hierarchical feudal order and as a set of values is known as *humanism* (Hayles 1999).

Similarly, Machievelli's early 16th century work *The Prince* (Machievelli 2014) highlights the rise of the individual, a person capable of subterfuge and able to scale the previously unassailable walls of medieval order. The Machievellian individual has a strong and complex inner life, a psychological dimension that is self-willed and self-driven. This is the dark side of Renaissance man; this is the other side of humanism, where humanism, as defined in Thomas More's *Utopia* (More 2010) showed a regard and consideration for the plight of the individual with early projections of social planning. The emphasis is on the individual mind and what it can produce with the information of the new media, the book. The humanist, both good and bad, emphasizes the creative ability of the individual, which is still a dominant from a western perspective, and manifests in inventions like the *personal computer* (PC) focusing on individual use.

Further, Shakespeare's plays hold within them the archetypes of this new-self willed, mobile reading and writing individual. A number of his tragic characters fail in part because of their inability to understand and operate the new media of the day: the book. Macbeth, who does both read and write in the play (Shakespeare [1623] 1980), fails to *interpret* the riddle of the witches, showing himself to be a poor reader. It is a tragic irony that this man who can both read and write is hasty and unreflective when listening to the prophecy of the witches, believing in a single interpretation of their message—that he cannot fail in his endeavors—propelling him into a sequence of murders which do not secure him his desired crown. Hamlet, on the other hand thinks extensively about the philosophical outcomes of life, with stage directions showing him reading a book (Shakespeare [1623] 1985). But his sophisticated thoughts of life and death as typified in his soliloquy 'To be or not to be....' (ibid.) hinder him, as he fails then to secure a successful revenge of his father's murder. Hamlet is an over-reader, unable to balance thought and action, not understanding how the new media of his day should fit into his world and serve his political ambitions. Whilst in *Julius Caesar* (Shakespeare, [1623] 2004) it is Mark Antony, an ancient orator, who holds an unopened scroll—very, very old media—on which Caesar's will is written. He holds this and speaks to the crowd, emphasizing the murdered Cesar's generosity but uses the presence of the unread document dramatically to reinforce a promise and thus swings political favor in his direction, finally cornering Brutus and the fellow murderers. Shakespeare has shown his audiences how the new media of the day can be used successfully or unsuccessfully to support individual ambition.

Psychology, a late twentieth century invention, can then be understood as the end point of the age of the individual (Hughes 1981): psychology discovers, describes and measures the interiority of the individual mind. Psychology shows how the individual is constructed, operates, and is motivated to fulfill individual projects and is guided by old media. The literary technique of *stream of consciousness*, championed by James Joyce in *Ulysses* (Joyce 1922) and Virginia Woolf in *Mrs. Dalloway* (Woolf 1925), signaled a new dimension and perspective of the individual in literature where we, the reader, were inside the character's head, so to speak. It is no mistake that Freud wrote extensively about Shakespeare's *Hamlet* (Freud 1899). The question is then, what happens to 'mind' and to body as an inherited notion of the individual, when new media are introduced, namely the computer?

3   Painting: A Medium with a Message that Reflects upon the Medium

Closer to the age of computing in the mid twentieth century, painting has offered intelligent discourse on how we see and experience the world. Painting is an innovative medium, creating new ways of seeing where the old vision is replaced by the new. That old media are absorbed by new media as McLuhan has stated (McLuhan 1964), is also true for the aesthetics of painting. What follows is a brief commentary of significant art over the last century with an initial reference to the Renaissance, commenting on the role that machines have in the production of art and in human perception potentially offering a model for interactive film and computing.

Concurrently, in the Renaissance there is the new media *image* that emerges on a parallel path with the book. That which is seen is recorded in two dimensions via perspective drawing, the converging lines delivering the illusion of space on a flat plane. The evidence of this technology as a system is first dated as 1350 with artists such as Albrecht Dürer later using grid screens in front of their subjects to map by 1525 (Manovich 2001). It takes some time for an understanding that this technology can depict real space rather than just mythical narrative space. The new technology of perspective drawing has its parallel to the age of Shakespeare in the Dutch masters, painting depicted 'reality' with a soft light. Not only is sound recorded in the form of the book but so are the effects of light. Again, we see that is the nature of the medium—its capacity to deliver certain types of messages—that has particular effects. Form and content are closely aligned and in this case painting delivers the seen as *reality*.

Da Vinci's image of man known as *Vitruvian Man* (Floch Le-Prigent 2008) (see Fig. 12.1) a figure drawn with outstretched hands, a figure depicted realistically, yet with a sensitivity of observation that is subtle and unique, characterizes the visual representation of Humanism.

The image at once idealizes the human form and places it clearly in view. The measurements and the dimensions resonate and systematize an aesthetic of the human. The realistic depiction is as Baudrillard describes 'seductive' (Baudrillard 1984) in that we cannot see past it readily; the illusion of perspective as reality is potent, communicating the content of the new man. Man here is powerful yet sacred, vulnerable and sensitive in his earthly form, measured and outstretched, a naked creature, on the Humanist pedestal.

To then jump some four hundred years we see in the work of Picasso's and Braques's Cubism (Hughes, ibid. 1981) a world dramatically changed by the new machines of the early Twentieth Century. At this time, a time when industrialization is in full swing with the motor car, electricity, flight and the growth of massive industrialized cities, human experience was forced into a speed and complexity of experience that was entirely novel. What Cubism offered was an instantaneous view of the complexity of human experience; life was now not a matter of a single beautiful aesthetic view to be painted, but rather a clash of different sometimes jarring and conflicting views into one. At the same time what Cubism was resonating with the new science of psychology, which both Freud and Jung were evolving at the time, the interiority of the individual. Cubism also resonated with new viewpoints in science, namely Einstein's theory of Relativity, supporting the notion that knowledge evolved from the single privileged observer was a limited view of how reality was constructed. Cubism, in this way, was a medium with a new message and expressing the new complexity of the individual.

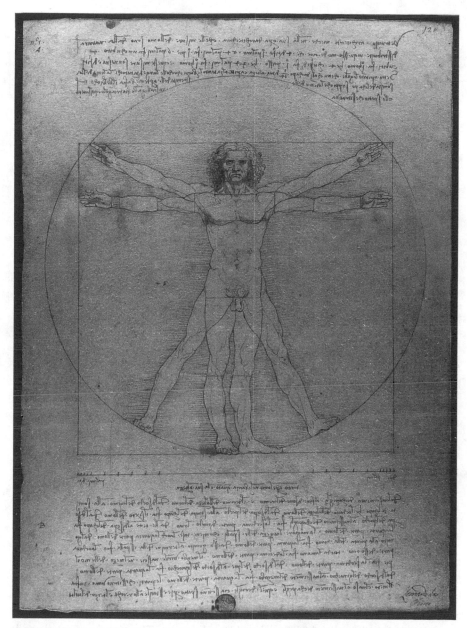

Fig. 12.1 Vitruvian Man: the human form idealized (circa 1490)

Fig. 12.2 Fountain 1917,
Duchamp

4 Photography and Modern Sculpture: Machine-Made Art

Painting, a light based medium, is however subverted by photography (Benjamin 1955). The process of manual mimetic depiction is suddenly automated and a window on life as exact simulated reality is made. The new image making of photography made redundant the craft of drawing and realistic painting obsolete. Photography spread rapidly and became popular and was made available to a mass consumer public. The new here had also absorbed the old medium. The same was true for magnetically recorded sound developed by Edison; the word of text in one swoop is made redundant by magnetic sound. When sound and light converge in the form of film (Manovich, ibid. 2004), with the understanding that successive photographic frames delivers an illusion of movement, just as perspective delivered the illusion of space, then the process of mimesis, or reproducing reality as we see it, is complete in the form of film. At this time also Duchamp, with the exhibition of the pissoir at a Paris exhibition in 1917 (Hughes, ibid. 1981) and giving it the title *Fountain,* was announcing the redundancy of the old media in art. The pissoir, an ironic and cheeky object to submit as an art object, is of course industrially made by machine after an original form has been made by a single craftsman Fig. (see Fig. 12.2). Duchamp was commenting wryly on the domination of the machine, that the artisan, the maker by hand was being made redundant by industrial processes.

It was Warhol, in the twentieth Century, however, who suggested that not all was lost in the land of art as a result of machine-made art. Instead of decrying the loss of painting, he invented a visual aesthetic in his work, the aesthetic of mass produced

Fig. 12.3 "Mask II" by Ron Mueck. (Courtesy of Jack 1856 on Wikipedia)

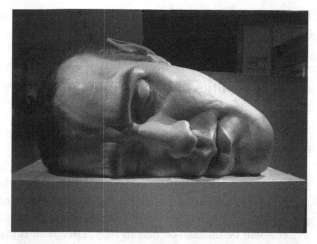

images. Warhol emphasized that *repetition* was an aesthetic principle, which indeed it is. In his work of Elvis Presley pulling a gun, it is the multiple representations of the drama of pulling a gun—an extreme act—that highlights the fact that the message is change by the repetition, its impact lessened and modified by the machine reproduction. Whilst in his multiple works of Marilyn Monroe, the changes in chromatic key of the woman most popularly known in the mass media of the day as the most beautiful of all women, indicated that the new palette and brush of the artist was push button color selection, replacing brush and canvas.

Richter then follows Warhol, a popular contemporary artist of the late twentieth and early twenty-first century. His approach was to develop an aesthetic of an image-producing machine that malfunctions, a machine that like Warhol's produces images but produces not as designed or ordered. The work might be the refuse of any common color-copying machine in any office in the world. His work is on canvas and is handmade, creating an irony from the early departures of Duchamp.

Whilst akin to painting is the sculpture of Ron Mueck, a contemporary Australian artist who was rejected many times from even studying art at contemporary art schools. Mueck's work emphasized scale and hyperrealism as aesthetic concepts (see Fig. 12.3). The work is of interest in this discussion because the computer similarly has an exceptional capacity for realism and whilst also managing scales of data not previously available to users. The sculptures beckon the questions not so much of aesthetic appreciation—is beautiful, is it ugly, rather provokes the audience into questions of relationships; what is my relative understand of this work? How should I respond to realities that are unusual and outside the scope of my experience? A shift in scale has direct relevance to computing as experienced with the advent of Big Data.

In summary, in the artworks we see a strong and well-developed habit of artists using their medium in ways that often reviews and comments upon the limits and potentials of the medium as perception and self-perception. The artworks discussed show an understanding of how a medium may be superseded by a newer, more technically or intellectually evolved medium. Many of the artists discussed addressed the problem by repositioning or re-tasking the old media meaning that new perspectives on the new tools emerged. In short, they asked what the medium was in its new form and shaped messages to that medium accordingly.

5   The Qualities of Interactive Film

> ....electronic space becomes a medium of action rather than information
> Stelarc (Donald 2003)

The approach here in examining medium and message is not so much philosophical as material: how does the medium perform and what can it communicate taking into account the impact of media and their messages as evidenced in past epochs? A single distinction is made here, in that the depiction of on-screen images by computer systems will be understood here as *interactive film*: film, because of the pervasive real and animated images; the motion; and interactive, because these images change with the addition of any data through exchange of information with databases.

To look at any personal computer or smartphone it is in plain view that the computer is a machine that makes other machines (Crogan 2008); it absorbs older forms of media be it text, images, sound and simulates their real functions virtually. The notion is quite useful and in part is responsible for the difficulty in being able to define its characteristics as a medium making messages because of the scale and complexity. The singular, object associated nature of media and media machines is historical and physical; printing presses, cameras, typewriters, microphones etc. and so on are knowable and have stand alone functions. The explosion of what are known as 'apps' in mobile computing, generating $ 200 billion of business in 2004 (Varshney 2000), is the result of this 'machine making other machines' capacity. Hence, rapid proliferation and a change of scale into million if not billion of interactions and more, is a central feature of the medium and its messages.

Interactive film, which is what we see as the computer screen surface, is then the result of convergence. The word 'converge' suggests that things are becoming focused into a single almost absolute point. The real underlying principle of interactive film is a marriage between film and programmed interaction. Hence this compounding of the two processes, one the simulation of a material process, the other electronic—brings about the invention of screen representations whilst also referring to a database. The *telegraph* is also a convergence; electrical conductivity was converged with the intellectual/physical process of speech. The notion that the current of a wire could be interrupted and that the interruption of the current could follow a code representing the sonic alphabet was a marriage, a convergence.

Essentially there are three main characteristics that support calling the medium of the computer, interactive film:

The first is the **presence of touch** as a predominant interaction mode (Danylak and Edmonds 2006). If we take the example of the ever-popular smart phone and tablet interface now in extensive use, the management of the user applications is through a gentle left and right motion finger touch motion. Clearly this resembles film in its early manifestation travelling through an edit machine (see Fig. 12.4 and 12.5).

Fig. 12.4 Manual film editing

Fig. 12.5 Smartphone simulating film surface

The current success of smartphone technology is the result of adopting a film metaphor by designers. The design of the surface is useful from a user perspective in that the volume of information for the user is limited by the left and right movement. A constant difficulty for the designer is the appearance of overwhelming amounts of data representations; for the user, the visualization of user functions must be coherent and not lead to cognitive overload. The adoption of the film as a material design metaphor has solved this pervasive and difficult problem with astounding commercial success resulting.

The **second is convergence of the database with the film surface**. The constant movement of the film surface to and from the database is a unique feature of interactive film. This is best understood from the perspective of *interaction*; we can say that each viewing of a screen presents us with a new frame. Film traditionally operates as a medium travelling in a single playback direction delivering the illusion of movement at 26 frames per second. The computer readily delivers the playback of digital film copying its analogue form. The other depictions of media forms are also film projections but they are often still frames. Rather than travelling along the axis of time, the graphic user interface (GUI) refreshes the image in the database, adding or subtracting data as directed. In this way the refreshed screen can be understood as a film edit, or a new frame in the original experience of the screen. A significant difference in the traditionally fixed film playback and interactive film is that fixed film is repeatable and passive. Interactive film may be repeatable in its original form and unaltered by interaction and commands sent to databases makes the relationships between the interactive entirely film unique (Danylak and Edmonds 2012).

The **third is the nature of representations** classified as either realistic or animated, a distinction that is typical of film (Kelty and Landecker 2004). The photographic basis of film allows itself to make either real or animated depictions.

The real, is that which has an apparent parallel from the image to world which we share and experience; animation is the capacity to draw or fabricate illustrations which have life-like qualities also based on a photographic process. A look at any smartphone and the representations of its applications indicates a strong mix of both the real and animated. The distinction often goes undetected.

6 Conclusion

The focus of this chapter has been to briefly survey media historically up to and including the computer and how it absorbs old media and to describe the characteristics of new media. One modern medium stands out in the process and that has been the medium of film, and as discussed, it is interactive film—the convergence of film with database interaction—that best describes the medium of emerging human computer interaction.

From the viewpoint of the emergence and socialization of the book, we can see that a type of person, the individual, was not only the content—the message—of the medium, as featured in Shakespearean texts, but that the mechanical reproduction of the medium and invention of that time also encouraged individual interpretation to occur through increased access to information. Importantly, this is a manifestation of sound.

Notable comments regarding cultural adjustments to these changes have been expressed by Hayles (Hayles ibid.) and by Weiner (Weiner 1954). Hayle describes our current epoch as the *posthuman*. Her thesis is that we have left the age of humanism and that the unclear distinction, a symbiosis, between human and machine is now in play resulting in a time that is after the humanist epoch but still leans heavily upon it. Weiner, who coined the word 'cyborg' deriving it from the Greek work *gubernator* meaning 'helmsman'—the one who steers a boat through its journey—described the human as a 'a flame', in that all that is human could be defined as information, patterns that are ceaselessly in motion. The thesis proposes that information stands between matter and energy and is the vital and essential link between the two.

It is noteworthy that painting has been a medium that has a highly functional and practical beginning in terms of perspective drawing. Its evolution however has gone far beyond that, displaying a capacity to self reflect in ways that responded to other media technologies and to crises of redundancy in the art form, creating innovation within the medium. The capacity of painting to do so is very interesting and may explain its long-standing success to take a dominant role culturally for so many centuries. Importantly, this is a manifestation of light.

In the case of interactive film we are still only just beginning to understand the behavior of the medium. The three qualities—transport, interaction, and the dual representation of the real and the animated—are highly significant factors in the current success of the computer. The summary here has been focused more on the medium rather than the message, a result of the still overwhelming development and invention of the technical aspects of computing. Importantly it is the recording of both sound and light that are at the foundation of film.

Acknowledgement The chapter includes sections of a paper Danylak R. 2014 Interactive Film: the Computer as Medium, HCII 2014 Crete (Springer-Verlag).

References

Andersen, P. B., 2001, 'What Semiotics Can and Cannot Do for HCI', *Knowledge-Based Systems*, vol. 14, no. 8, 419–424.

Baudrillard, J., 1984, *The Evil Demon of Images*, Power Institute Publications, Sydney, 14.

Benjamin, W., *Art in the Age of Mechanical Reproduction* in Illuminations: Selected Writings, 1955 (in German).

Crogan, P. 2008. Games, simulations and serious fun: An interview with Espen Arseth. *Scan: J. Media Arts Culture* 5, 1 (May 2008).

Danylak, R. & Edmonds, E., 'Touch as the act of signification; naming as a key design concept for gesturally intuitive interactive space', in *Engage: Interaction, Art and Audience Experience*, CCS Press, University of Technology, Sydney, pp. 62–67. 2006.

Danylak, R. and Edmonds, E. (2012) The planning and experience of time and space in three gestural media: theatre, film and interactive film, Int. J. Arts and Technology, Vol. 5, No. 1, pp. 1–16.

Dictionary, *Electronic Oxford American Dictionary* Version 1.0.2, Software, 2005.

Dijk, J. V., 2004, 'Digital Media', J. D. H. Downing (ed.), *Sage Handbook of Media Studies*, California, USA.

Donald, J., 2003, *Flannnery, Botanising the Interface* viewed 20.03.2006. http://www.icinema. unsw.edu.au/paper.html.

Floch Le-Prigent, P., *The Vitruvian Man: an anatomical drawing for proportions by Leonardo Da Vinci*. Morphologie. 2008, Dec 92 (299): 204–209.

Freud, S., *The Interpretation of Dreams*, Franz Deuticke, Vienna, 1899.

Hayles, Katherine, N., *How We Became Posthuman* University of Chicago Press, 1999.

Hayles, *ibid.*

Hughes, R., *Shock of the New*, New York, Knopf, 1981.

Hughes, R., *ibid* 1981.

Ibid. Act III sc. iii.

Joyce, J., *Ulysses*, S. Beach, Paris, 1922.

Kelty, C., Landecker, H., 'A Theory of Animation: Cells, L-systems, and Film'. *Grey Room*, Fall 2004, Vol. -, No., pp. 30–63.

Machievelli, N., *The Prince and other writings*, Canterbury Classics Baker and Taylor, 2014.

Manovich, L., *The Language of New Media*, MIT, 2001.

Manovich, L. *ibid* 2004.

McLuhan, M. *Understanding Media*, McGraw Hill, Toronto, 1964.

McLuhan, M. & McLuhan, E. 1988, *Laws of Media; the New Science*, University of Toronto Press, p. 7.

More, T., *Utopia*, Broadview Press, 2010.

Shakespeare, W., *The Tragedy of Macbeth* ed. A.N. Reimer University of Sydney, 1980.

Shakespeare, W., *Hamlet*, ed, A. N Reimer University of Sydney, 1985.

Shakespeare, W., Mouston, D., First Folio 1623 Facsimilie, Applause New York, 1995.

Shakespeare, W., *Julius Caesar* ed. Spevack, J., New Cambridge Shakespeare, Act 3 sc.ii, 2004.

Varshney, U., 'Mobile Commerce: a new frontier'. *Computer* Vol. 33, Issue 10 Oct 2000 pp. 32–38 USA.

Weiner, N., *The Human use of Human Beings*, Houghton Miflin, Boston, 1954.

Woolf, V., *Mrs Dalloway*, Hogarth Press, United Kingdom, 1925.

Appendix

Appendix A. Digital Artwork by Aleksei Kostyuk

Aleksei Kostyuk (visio-art.de), also known as VISIO, is an art director and digital artist in Munich, Germany. He has been gaining experience in the latter since he discovered his passion for Photoshop during an internship in a creative agency when he was 14 years old. In 2010, he graduated from the Macromedia Academy in Munich, majoring in graphic design. His creations have been published in books and magazines including Advanced Photoshop, Digital Arts, InPrint, and Pixel Arts. He is a member of three international art collectives: the Luminarium, Intrinsic Nature, and Slashthree.

Kostyuk portrays the step-by-step creation process for "Pump Up the Color" (see Fig. 1a–i) and "Scale of Life" (see Figs. 2a–h). "Pump Up the Color" was created for the Fotolia TEN Collection Contest; and "Scale of Life" for the 17th Exhibit of the Luminarium, an international art group with talented artists in both the digital and traditional realm from all over the world to pursuing originality and creativity.

Kostyuk's "iMobile" in Fig. 3a won the first prize in the mobile category at the Fotolia TEN Collection Contest; and his "Party Animals" in Fig. 4a won the Daily Deviation Award on Deviantart. Figure 3b–f and Fig. 4b–f provide a close-up look at the fine details of his digital artwork.

N. Lee (ed.), *Digital Da Vinci,* DOI 10.1007/978-1-4939-0965-0,
©Springer Science+Business Media New York 2014

Fig. 1 a–i Pump Up the Color. (Courtesy of Aleksei Kostyuk (visio-art.de))

Fig. 1 (continued)

Fig. 1 (continued)

Fig. 1 (continued)

Fig. 1 (continued)

Fig. 2 a–g Scale of Life. (Courtesy of Aleksei Kostyuk (visio-art.de))

Fig. 2 (continued)

Fig. 2 (continued)

Fig. 2 (continued)

Fig. 2 (continued)

Fig. 2 (continued)

Fig. 2 (continued)

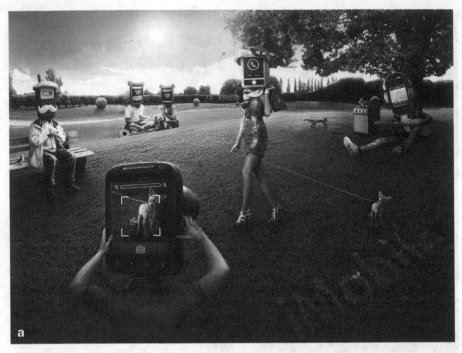

Fig. 3 **a–f** iMobile. (Courtesy of Aleksei Kostyuk (visio-art.de))

Fig. 3 (continued)

Fig. 3 (continued)

Fig. 3 (continued)

Fig. 3 (continued)

Fig. 3 (continued)

Fig. 4 a–f Party Animals. (Courtesy of Aleksei Kostyuk (visio-art.de))

Fig. 4 (continued)

Fig. 4 (continued)

Fig. 4 (continued)

Appendix B. Digital Artwork by Pawel Nolbert

Pawel Nolbert (www.nolbert.com) is a 1984 born multidisciplinary visual artist focused on working in the fields of illustration, art direction and graphic design. His work and techniques spans across many different mediums and formats as he has worked for a diverse group of internationally acclaimed clients on many different commercial and art projects. He has been working in the creative fields since 2002. His creative alter-ego "Hellocolor" and his work have been featured in a number of specialized publications online and offline, such as Taschen "Illustration Now 3", Computer Arts Magazine, Advanced Photoshop Magazine, IdN Magazine, or Adweek Talent 100 (2012).

Figures 5 and 6 show two digital images of Nolbert's ongoing "Atypical Gestures" project. "Art Direction" in Fig. 7 is an illustration ambiguously depicting art, art directors, designers, and the creative industry. "Offf—Year Zero" in Fig. 8 was a theme of the 2011 edition of OFFF—a post-digital culture festival hosted in Barcelona, the birthplace of the festival. Nolbert wrote about the piece, "The theme leads us to keywords like forgetting the past, reinventing ourselves, and setting up new rules. My piece 'Tabula Rasa' was a part of Ars Thanea presentation at OFFF and has been featured in the OFFF Year Zero book, along with pieces from other speakers and artists." Figure 9 is Nolbert's self-portrait illustration created for Adobe's "I Am The New Creative" Campaign.

Fig. 5 The number 8 in "Atypical Gestures." (Courtesy of Pawel Nolbert. www.nolbert.com)

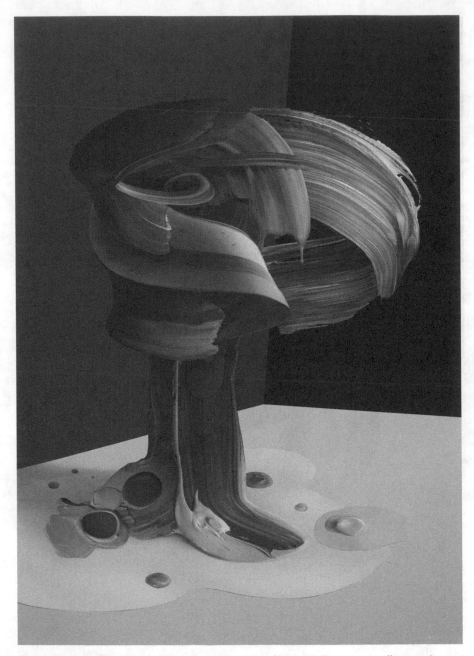

Fig. 6 The letter P in "Atypical Gestures." (Courtesy of Pawel Nolbert. www.nolbert.com)

Fig. 7 "Art Direction." (Courtesy of Pawel Nolbert. www.nolbert.com)

Fig. 8 "Offf—Year Zero." (Courtesy of Pawel Nolbert. www.nolbert.com)

Fig. 9 A Self-Portrait Illustration created for Adobe's "I Am The New Creative" campaign. (Courtesy of Pawel Nolbert. www.nolbert.com)

Appendix C. Digital Artwork by Tanya Varga

Tanya Varga (www.cellesriaart.com), formerly Tanya Wheeler, is a freelance artist based in the California Bay Area who works with both digital and traditional mediums. She attended the Academy of Art University where she studied 2D Illustration/Animation and Advertising. She currently works as a freelance graphics artist, concept artist and illustrator. She also does oil portrait commissions and has worked with traditional mediums, such as oil and charcoal, most of her life and discovered digital painting a few years ago and has loved it since. She is completely self-taught when it comes to digital painting and has picked up her experience from a lot of experimentation and practice. What she loves most about art is that it gives her the chance to bring visions to life.

Tanya's artwork has appeared in various publications including Spectrum (The Best in Contemporary Fantastic Art), Expose (The finest digital art in the known universe), Exotique (The world's most beautiful CG art) as well as Corel Painter, and ImagineFX among others. The examples in the following subsections are copyrighted by Tanya Varga and are reproduced in this book with her permission.

C.1 Digital Gallery of "Another World in the Universe"

Figures 10–16 demonstrate Tanya Varga's imagination of another world in the universe. She is often inspired and fascinated by the universe and the possibilities of finding other Earth-like worlds.

For the piece "Pink Super Earth Xianthen-18" (see Fig. 10), Varga wanted to create her own Earth-like planet. This imaginative world is "three times the mass of Earth and is surrounded by rings of diamond-rich asteroids. The four moons create enormous tides in its seas of liquid argon. The argon atmosphere ignites with light caused by surges of radiation from a nearby pulsar. If this planet could support any form of life, they would likely be very flat and luminescent from the electrified atmosphere."[1]

As a being that can create her own Universe of Dreams (see Fig. 11), she imagines the Light Sorceress (see Fig. 12), the Dark Sorceress (see Fig. 13), Allure (see Fig. 14), and the Curse (see Fig. 15), and the Artist (see Fig. 16).

---

[1] See Tanya Varga's Digital Gallery. *Cellesria.* http://www.cellesriaart.com/digital-gallery/.

Fig. 10 Pink Super Earth Xianthen-18 (2011). (Courtesy of Tanya Varga. www.cellesriaart.com)

Fig. 11 Universe of Dreams (2010). (Courtesy of Tanya Varga. www.cellesriaart.com)

Fig. 12 Light Sorceress (2011). (Courtesy of Tanya Varga. www.cellesriaart.com)

Fig. 13 The dark Sorceress
(2011). (Courtesy of Tanya
Varga. www.cellesriaart.com)

Fig. 14 Allure (2011). (Courtesy of Tanya Varga. www.cellesriaart.com)

Fig. 15 The Curse (2011).
(Courtesy of Tanya Varga.
www.cellesriaart.com)

Fig. 16 The Artist (2010).
(Courtesy of Tanya Varga.
www.cellesriaart.com)

C.2 Allison Harvard Inspired Portrait Speed Painting

Uploaded to YouTube in January 2012 (www.youtube.com/watch?v=D_jUV9x-PQg0), Tanya Varga demonstrated a quick portrait speed painting inspired by a model and artist named Allison Harvard who appeared on the TV show "America's Next Top Model" and who is well known on the Internet as Creepy Chan.

Figure 17 shows the finished image of the speed painting. Varga wrote, "Allison Harvard has a fascination with hemophilia so I thought the blood was a nice touch. I would also like to note that this portrait is merely inspired by Allison and was not meant to portray a perfect likeness of her. Her beautiful and unique features/personality only inspired me to create this portrait. I looked at a few images of her for reference to study her features but this image was not copied from any content. It was painted mostly in Corel Painter with some additional work done in Adobe Photoshop. Around 4 h of work, I used Corel Painter's symmetry tool for the first time to create this work. It's pretty useful for painting full frontal portraits and symmetrical designs. I stopped using it near the end though since I didn't want the image to be too symmetrical. Some of my brushes that I used in Photoshop are available at www.cellesriaart.com/resources/basic-brush-set."[2] Figure 18a–h highlight some of the key frames from the speed painting video to portray the amazing transformation from an empty canvas to a stunning digital portrait.

---

[2] See "Allison Harvard inspired portrait speed painting" by Tanya Varga on *YouTube*: www.youtube. com/watch?v=D_jUV9xPQg0.

Fig. 17 Allison Harvard
inspired portrait (2012).
(Courtesy of Tanya Varga.
www.cellesriaart.com)

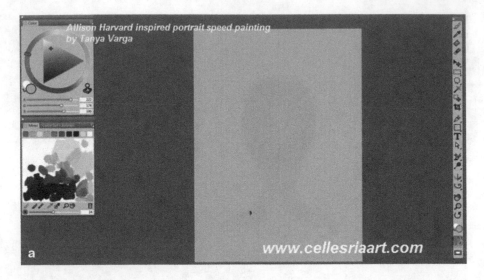

Fig. 18 a–h Allison Harvard inspired Portrait Speed Painting (2012). (Courtesy of Tanya Varga. www.cellesriaart.com)

Fig. 18 (continued)

Fig. 18 (continued)

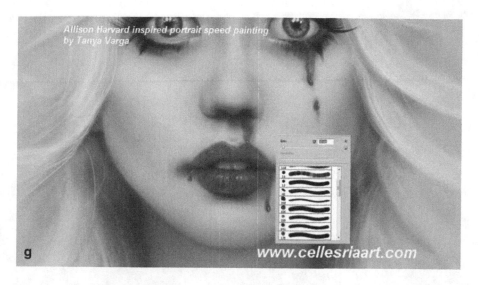

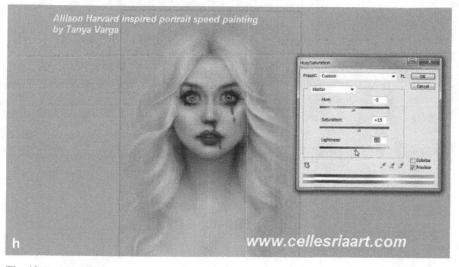

Fig. 18 (continued)

Index

Printed in the United States
By Bookmasters